Workshop Processes
First Level

Hutchinson TEC texts

Learning by Objectives
A Teachers' Guide
A. D. Carroll, J. E. Duggan & R. Etchells

Engineering Drawing and Communication
First Level
P. Collier & R. Wilson

Physical Science
First Level
A. D. Carroll, J. E. Duggan & R. Etchells

Workshop Processes and Materials
First Level
P. Collier & B. Parkinson

Electronics
Second Level
G. Billups & M. T. Sampson

Engineering Science
Second Level
D. Tipler, A. D. Carroll & R. Etchells

Mathematics
Second Level
A. Hill & G. W. Allan

Electronics
Third Level
G. Billups & M. T. Sampson

Workshop Processes and Materials

First Level

P. Collier & B. Parkinson

Hutchinson
London Melbourne Sydney Auckland Johannesburg

Hutchinson & Co. (Publishers) Ltd

An imprint of the Hutchinson Publishing Group

24 Highbury Crescent, London N5 1RX

Hutchinson Group (Australia) Pty Ltd
30-32 Cremorne Street, Richmond South, Victoria 3121
PO Box 151, Broadway, New South Wales 2007

Hutchinson Group (NZ) Ltd
32-34 View Road, PO Box 40-086, Glenfield, Auckland 10

Hutchinson Group (SA) (Pty) Ltd
PO Box 337, Bergvlei 2021, South Africa

First published 1980

Set in VIP Times Roman by D. P. Media Ltd

Illustrations drawn by Oxford Illustrators Ltd

Printed in Great Britain by The Anchor Press Ltd
and bound by Wm Brendon & Son Ltd,
both of Tiptree, Essex

British Library Cataloguing in Publication Data
Collier, Peter
Workshop processes and materials, First
Level. - (Hutchinson TEC texts; 7).
1. Machine-shop practice
I. Title II. Parkinson, Brian
621.7′5 TJ1160

ISBN 0 09 140491 6

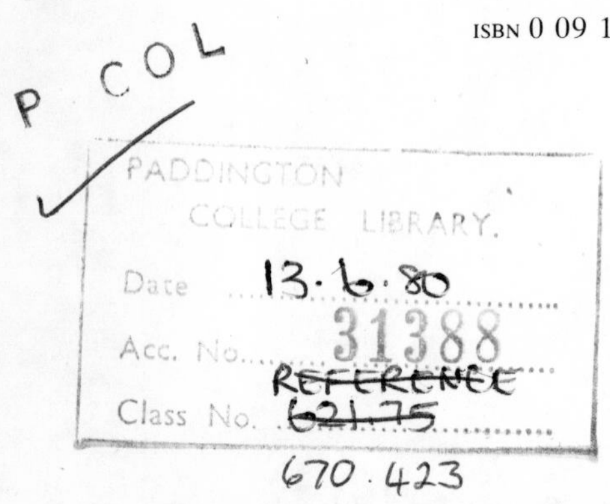

Contents

Acknowledgements

The authors and publishers wish to thank Dean, Smith & Grace Ltd for photographs of lathes; and Pratt Burnerd Ltd for photographs and details from which drawings of chucks were made.

Introduction

In each of the books in this series the authors have written text material to specified objectives. Test questions are provided to enable the reader to evaluate the objectives. The solutions or answers are given to all questions.

Topic area: Hazards in the workshop

After reading the following material, the reader shall:

1 Show an awareness of the hazards present in a workshop environment.

1.1 List the major responsibilities of the employee under the Health and Safety at Work etc. Act.

1.2 Identify the dangers associated with unsuitable clothing, hair and machine guards, when working with equipment which has moving parts.

1.3 State the common methods of stopping equipment in an emergency.

1.4 Explain the need for eye protection in relation to sparks, dust, etc.

1.5 Identify hazardous conditions in the workshop environment, given a workshop layout.

1.6 List the dangers associated with electricity in the workshop.

Safety is a word used with increased regularity, and yet an understanding (to any depth) of safety in a workshop is the exception rather than the rule. An accident in a workshop or machine shop may involve the loss of a limb, blindness, disfigurement or even death. When investigated, the cause can be equipment left lying in a gangway, loose clothing caught in a machine, a defective or incorrectly fitted guard, inefficient clamping, or a careless movement. With care, many tragic accidents could be avoided. Having the attitude 'It can never happen to me' is not a guarantee of safety.

A copy of the Factories Act can be found in all workshops and factories; the act sets standards that must be maintained by the employer and the employee. The Health and Safety at Work etc. Act 1974 states:

Duties of employees

Section 7: It shall be the duty of every employee while at work:

(a) to take reasonable care for the health and safety of himself and of other persons who may be affected by his acts or omissions at work;

(b) . . . to co-operate with his employer or his representative so far as is necessary to enable a duty or requirement to be performed or complied with.

Section 8: No person shall intentionally or recklessly interfere with or misuse anything provided in the interests of health, safety or welfare.

Thus, for an employee the essential features of the Act are:

1 The individual should take care of himself and his workmates.
2 Safety equipment must be used with respect, and any losses or damage must be reported.

Employers, by law, must provide certain items of equipment in the interests of safety – machine guards, goggles, helmets, etc. It is in the interests of the individual to be aware of the uses of such equipment. Safety equipment is rarely cheap and may, if broken or missing, jeopardize the life of a workmate. If accidental damage does occur, or some item is missing, this should be reported immediately.

Personal clothing

Figure 1 shows a form of dress suitable for use in a workshop.

Unsuitable clothing is a source of danger in a workshop. Overalls should be close fitting, with sleeves that either button at the wrist or are rolled up above the elbows. At no time should any loose clothing be worn; loose clothing includes such items as ties, scarves, torn sleeves or unbuttoned boiler suits. Serious accidents have resulted from the entanglement of loose clothing in revolving spindles, chucks and drills. Jewellery (e.g. rings and medallions) presents a similar problem and should not be worn. Unguarded long hair is dangerous when operating a machine. If it is caught in some moving part it may tear skin off your head. Special caps which enclose hair are provided and should be used. If there is a risk of damage to the head from work being carried out overhead, or from low-level structures, a safety helmet must be worn.

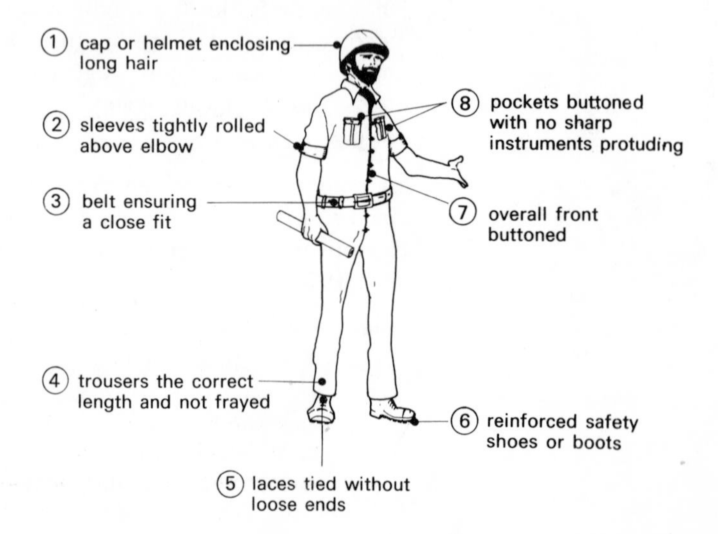

Figure 1 *Correct dress*

Figure 2 shows a reinforced industrial shoe. Shoes should give adequate protection both at the toe cap and at the sole; steel reinforced toecaps in boots and shoes that have thick soles can prevent a crushed foot or the loss of a toe.

Gloves must not be worn when operating machine tools; they may become entangled with the point of a drill or other moving parts, which could result in the loss of a thumb or a finger. However, when handling raw materials such as sheet, bar and castings, industrial gloves should be worn, reducing the risk of lacerations to the hands.

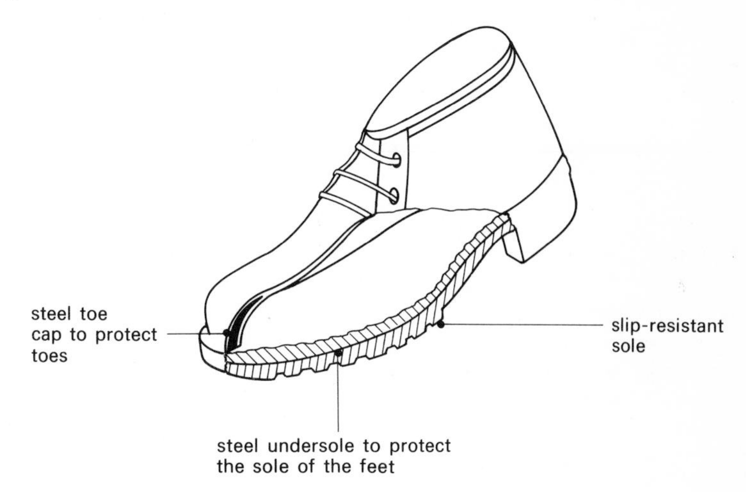

Figure 2 *Safety shoe*

Eye protection

The importance to an individual of sight is self-evident. Special care must be taken both by the employee and the employer to protect eyesight from injury in hazardous situations. The government, recognizing the need for special care, has published the 'Protection of Eyes Regulation'. The regulation states that, for specific purposes where there is a risk of injury to the eyes, eye protectors or shields must be worn at all times. The specific purposes where there is a risk of injury to eyes include any purpose which causes metal splinters, dust, grit or radiation to be produced in such a manner that injury to eyes could occur. In these conditions eye protection must be provided by the employer and must be worn by the employee.

Examples of eye protection equipment are shown in Figure 3. These are available with lenses which, in addition to protecting the eyes from injury caused by metal splinters or grit, also protect the eye from harmful radiation produced during electric welding operations. For those who require optical spectacles, special toughened optical lenses can be fitted.

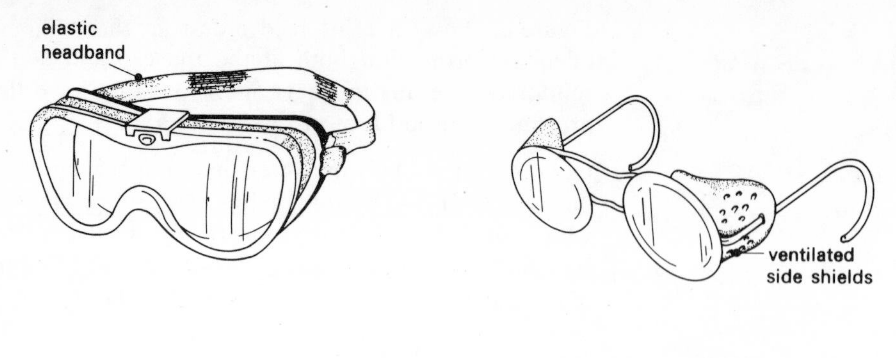

Figure 3 *Spectacle-type frames fitted with side shields and safety-glass 'lenses' for light duty*
Transparent plastic 'visors' can be worn over ordinary spectacles: they give a wider field of view than goggles, but are not so strong

Note that the employer must provide eye protection for use in hazardous situations, but it is the employee who is responsible for using them.

Workshop layout

Order and tidiness are essential in the workshop; all working areas should be maintained in a clean uncluttered condition. Floors should be clean; any spillage of oil or other liquids should be cleaned immediately, using an absorbent substance such as sawdust or sand.

The risk of accidents can be significantly reduced by good design of the layout of the workshop. All machinery should have sufficient space surrounding it so that the chance of people approaching too near rotating parts is negligible. Gangways must be kept clear at all times, and the boundary between the gangway and the working area clearly marked. Machines should be positioned so that no part encroaches into the gangway. Also, the operator of a machine should never have to stand in the gangway.

There should be appropriate areas for the storage of tools and equipment, and access to these areas should be simple.

There must be washing and toilet facilities, and provision for basic first aid. Again, these facilities should be easily accessible.

Machining with safety

Before any machinery is used, the operator should be completely familiar with all of its controls, particularly the location and operation of the emergency stop switch or button. (If a machine has to be stopped in an emergency, the electrical supply, rather than the mechanical supply, should be cut.)

The stop button must be easily accessible to the operator (on large machines this may mean more than one stop button), and must be easily found. The stop button is usually a mushroom-shaped knob, painted red. In contrast, the start button, usually painted green, should be shrouded so that the machine cannot be started accidentally.

In addition to the stop button on each machine, there should be several stop buttons at appropriate places in the workshop. Their function is to cut the electrical supply to ALL machines, rather than to an individual machine. They are needed on the occasions when an operator cannot take action to stop his machine.

Operators should also be familiar with methods of effective guarding against the dangers of revolving shafts, spindles, cutters and chucks. Figure 4 shows typical guards that are found in a modern workshop. Such guards offer protection to the operator's hands and eyes, and should always be used. Flying pieces of broken tools are a major cause of industrial injury.

Figure 5 shows several methods that may be used to clamp a workpiece to a machine table. In positioning a workpiece for machining, it is essential that it is securely clamped. This activity takes time and effort, but may prevent a serious accident.

Machinery should never be left running when it is not being used. Any measurements, adjustments or changes to settings should only be made when the machine is stationary. Swarf should be removed frequently, but only when the machine is not in operaton. Turnings or shavings should not be touched by bare hands. A piece of metal rod or other device should be used in order to reduce the risk of cuts and skin infections.

Cutting oils need replacing regularly, and care should be taken so that they do not come into contact with the skin. The use of barrier creams, in conjunction with the wearing of protective clothing, reduces the risk of infection.

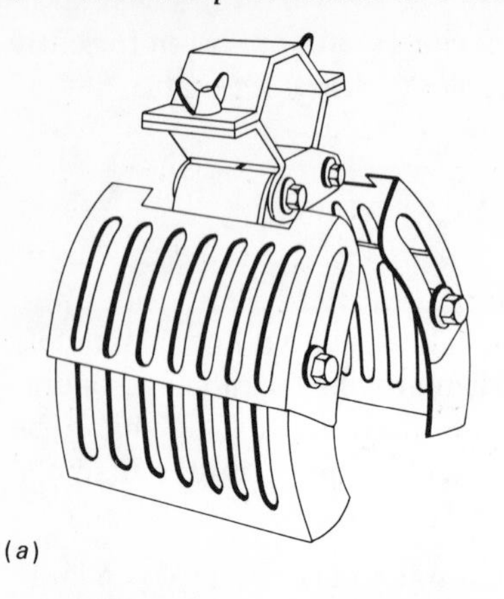

(a)

(b)

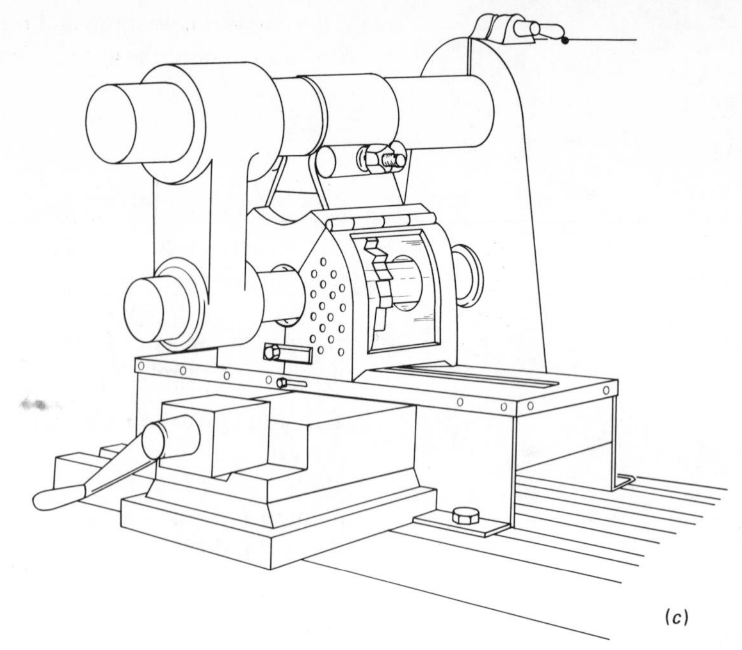

(c)

Figure 4
(a) *A guard comprising two telescopic members of transparent material which can be so adjusted that when the drill is at its topmost position, the lower member extends below the point of the drill. Being attached to the quill or sleeve, the guard moves up and down with the drill, collapsing as the drill enters the workpiece and extending when the drill is withdrawn. It is hinged to open sideways for drill changing and is retained in the closed position by a spring. The guard is particularly suited for small bench-drilling machines*
(b) *A proprietary guard which can be adjusted to suit cutters of varying diameters. When used with a false table, the guard affords a good standard of protection*
(c) *A good degree of protection is provided by this fixed guard and false table. A transparent panel in the guard permits observation of the milling operation*

Electrical safety

The effects of an electric shock on a person depend upon the magnitude of the current and the time that the person remains a part of a circuit. The magnitude of the current is determined by two factors:

(i) The potential difference across the circuit
(ii) The resistance (which depends upon the area in contact with the live 'metal', the condition of the person's skin and the length of the path taken by the current).

At 12 volts, some people are aware of a tingling sensation. Between 20 and 25 volts, many people feel a shock if the conductor is gripped

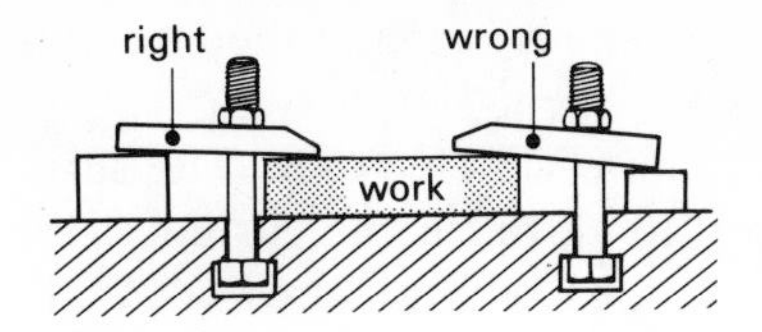

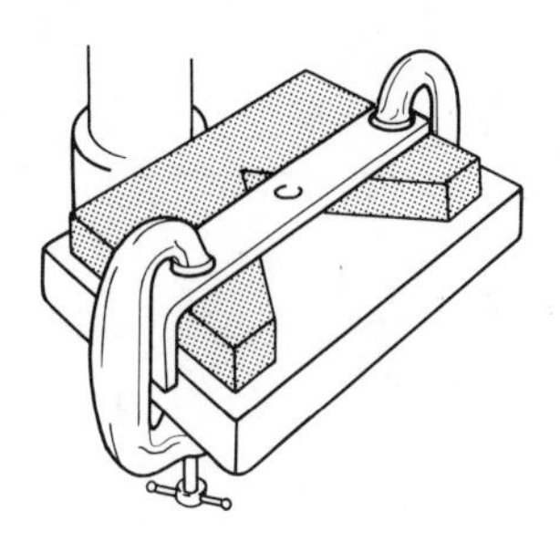

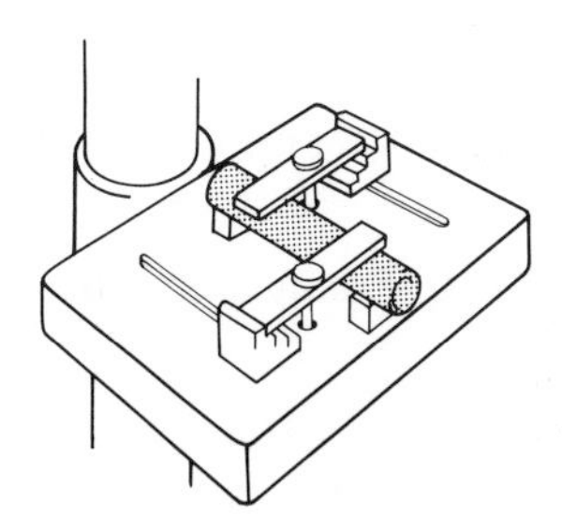

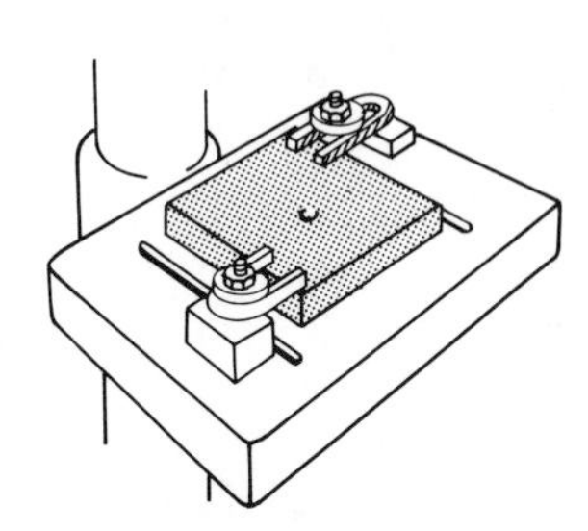

Figure 5 *Typical clamping methods*

firmly. When the potential difference (voltage) exceeds 25 volts, the muscles of the human body may contract involuntarily. This may cause the person to grip the conductor more firmly, resulting in the area of contact increasing. If this occurs, the electrical resistance is reduced, and for a constant voltage, an increase in current occurs. If the current is sufficiently large, breathing may cease or the heart may beat erratically. Death may occur; indeed death has occurred at a potential difference of 60 volts a.c.

Before attending to a person who has suffered an electric shock, assistance must be summoned. Then the main power supply must be switched off, or the rescuer may also suffer an electric shock. The victim should be given artificial respiration as quickly as possible, and medical assistance sought.

Electrical energy is supplied to appliances by means of conductors, which are insulated to prevent the leakage of current. Insulation materials tend to be mechanically weak; plastics are commonly used, and it is important that they be protected from any damage. A common method of protection is to shroud the cable in a metal tube or conduit, or a tough flexible sheath (Figure 6).

People and equipment can be protected against the effects of some electrical faults by 'earthing' the metallic non-current-carrying

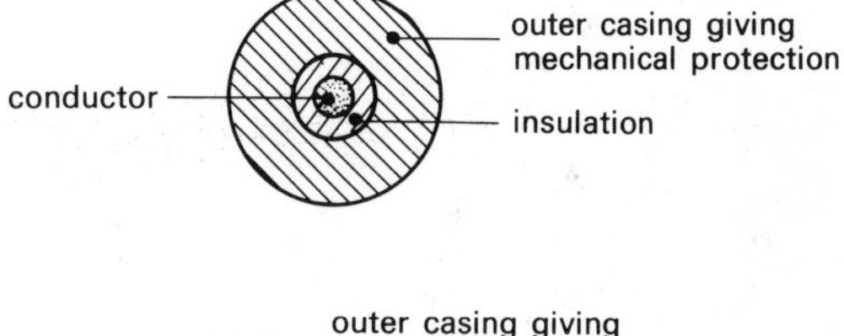

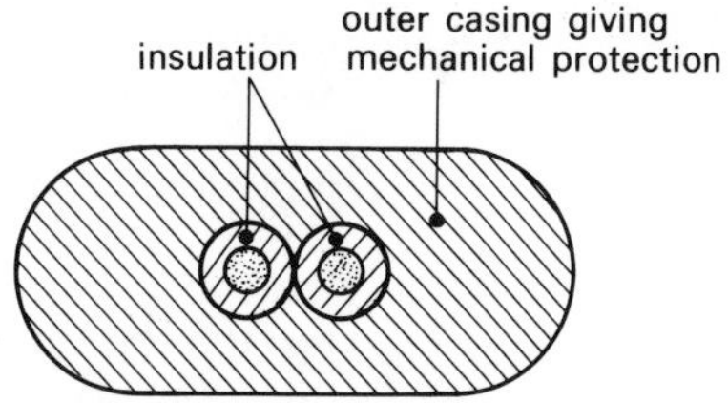

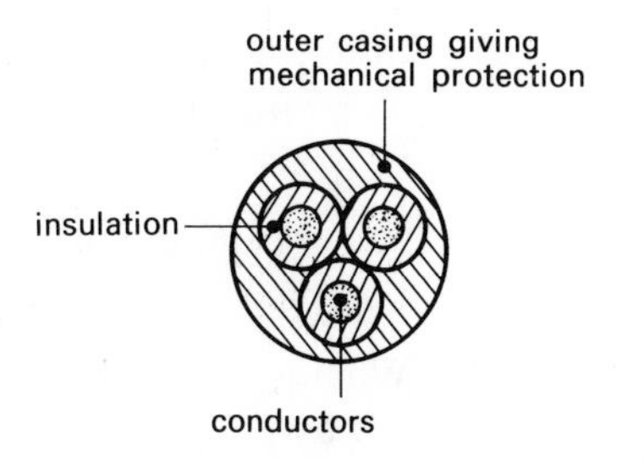

Figure 6 *Single-, double- and triple-core cables showing insulation and outer sheath*

frame of an electrical appliance. Earthing is accomplished by connecting all the metal (other than that designed to carry current) to an earth lead which is connected to an earth electrode. The earth lead provides a low-resistance path to earth for the current. This current may then operate safety devices such as fuses or circuit breakers, which interrupt the electrical supply.

Figure 7 *Personal clothing*

Self-assessment questions

1 List two of the main responsibilities of the employee under the Health and Safety at Work etc. Act, 1974.

2 State the main reasons why care should be taken with
(i) clothing
(ii) hair
(iii) shoes
(iv) gloves
in a workshop situation.

3 Figure 7 shows a workman whose dress includes eleven points that are a hazard to safety. List these hazards.

4 State why rings and medallions are sources of danger to a person working with rotating machinery.

5 State the purpose of using barrier creams in a workshop.

6 Explain why goggles must be worn when grinding.

7 Figure 8 illustrates a small workshop which has been poorly planned. Sketch a better layout for the workshop, using safety as the prime consideration.

8 In an emergency, a motor-driven lathe spindle should be stopped by isolating it from its *mechanical/electrical* supply. Which italic word is appropriate?

9 The colour of the stop button on a machine is red.
TRUE/FALSE

10 The stop button on a machine tool is shrouded so that the machine cannot be switched off accidentally.
TRUE/FALSE

11 In an emergency, it is possible to stop *all* of the machines in a section of a workshop by pressing one button.
TRUE/FALSE

12 Select from the diagrams in Figure 9 the most effective method of clamping.

13 What is the approximate voltage at which an electric shock can be felt by a person touching a live conductor?

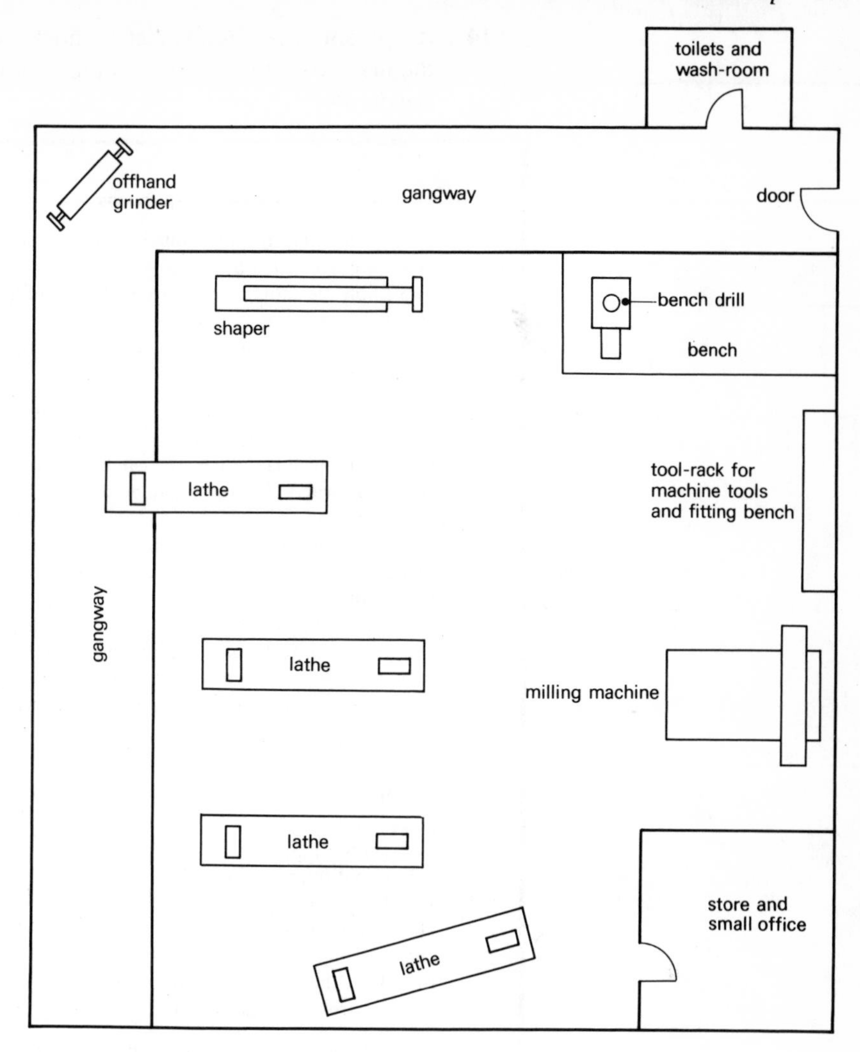

Figure 8 *Self-assessment question 7*

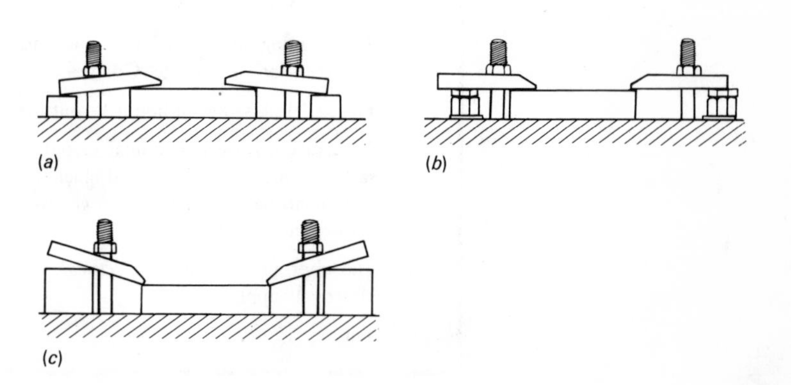

Figure 9 *Clamping alternatives for self-assessment question 12*

14 If a person receives an electric shock and is not conscious, what are the first two actions one must take upon arriving at the scene of the accident?

Solutions to self-assessment questions

1 Two of the main responsibilities of the employee under the Health and Safety at Work etc. Act 1974 are:

(i) The employee should take reasonable care of himself and anyone else who may be affected by his acts

(ii) The employee must use with care the safety equipment that is supplied, and should report any damage or loss

2 (i) Loose-fitting clothing or loose pieces of clothing such as frayed cuffs or scarves can be caught by moving machinery; serious injuries may result.

(ii) Long hair which is not covered can be caught by moving machinery (e.g. spindles, chucks or drills). The result can be that hair is pulled out, taking with it some skin from the head.

(iii) Shoes that have unprotected soles can result in the sole of a workman's foot being cut or pierced by sharp objects. Rubber or leather offer little protection against high temperatures; severe burns to the feet are possible. Standard shoes have uppers made from supple materials which provide little protection against heavy blows.

(iv) Gloves are in effect loose clothing, and may be caught in moving machine parts. They should not be worn when operating machine tools. When handling materials and tools, gloves should be worn to reduce the risk of injuries to the hands.

3 1 long hair left uncovered
2 overall front unbuttoned
3 torn breast pocket
4 frayed cuffs
5 trousers too long
6 lightweight unprotected shoes
7 shoe laces untied
8 hole in rule pocket
9 cuff left unbuttoned
10 overalls without a belt and left loose
11 sharp instruments sticking out of breast pocket

4 Rings or medallions may become entangled with rotating machinery. Damage to the person (and to the machinery) is then likely.

5 Barrier creams prevent oil, dirt and germs from entering the pores in the skin. Thus the risk of infection is reduced.

6 The grinding process produces dust and small grit particles. Both of these can cause eye irritation or damage. Goggles prevent the occurrence of such accidents.

7 Figure 10 illustrates a better layout for the workshop.

There are a variety of similar layouts, but the basic principles of them all are the same: sufficient space around machines; gangways left clear; machines positioned so that no part infringes upon gangways; clear access to the store, tool rack and washroom.

8 The appropriate word is *electrical*. A machine should be isolated by cutting its electrical supply.

9 True.

15 Why are the metallic frames of electrical appliances connected to earth by an earth conductor?

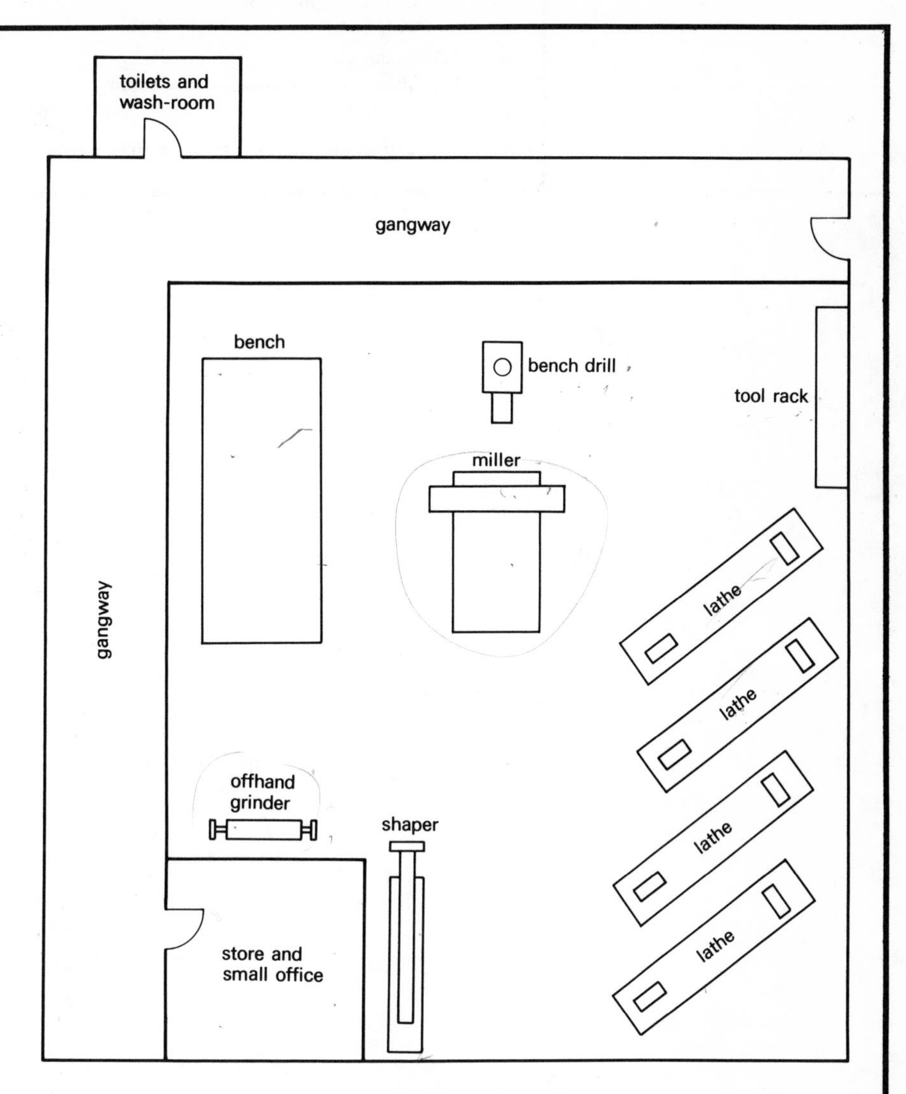

Figure 10 *Solution to self-assessment question 7*

10 False. The *start* button on a machine tool is shrouded so that it is unlikely to be started by accident. The *stop* button is unshrouded and easily accessible.

11 True.

12 The most effective method of clamping is that shown in diagram (c).

13 An electric shock can be felt at approximately 20 to 25 volts.

Answers to self-assessment questions

14 The first two actions are
(i) Summon assistance.
(ii) Isolate the victim from the supply by:
(a) Switching off the main electrical supply.
(b) Removing the victim by pulling on non-conducting material.

15 An earth conductor provides a path to earth for leakage current.

Topic area: Hand processes

After reading the following material, the reader shall:

2 Understand factors involved in the selection of hand tools for various tasks.

2.1 Select basic hand tools to perform a variety of tasks.

2.2 Select basic powered hand tools to perform a selection of tasks.

2.3 List the relative merits of powered and non-powered hand tools in terms of speed of production, cost, accuracy, fatigue.

In an engineering workshop, the processes of material removal and forming of materials are major activities. They are usually accomplished by machine tools, but occasions arise when it is necessary to use hand tools.

The fitter's vice

Material removal and forming requires forces to be applied in prescribed directions. The workpiece must be secured. One method is to use a fitter's vice, which is bolted firmly to a work-bench. The vice body consists of two main parts, the sliding jaw and the fixed jaw (items 1 and 2 in Figure 11); the material of both items is cast iron. Movement of the sliding jaw is by means of a mild steel screw, (item 3 in Figure 11), turned by a 'tommy bar' (item 4 in Figure 11).

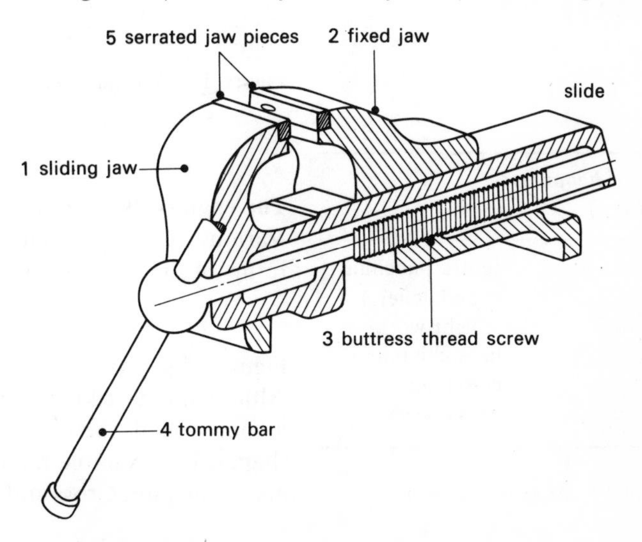

Figure 11 *Fitter's vice*

Figure 12 shows the directions of the constraints on movement of the workpiece. The symbols 'a', 'b' and 'c' show restraint to linear movement, whilst 'd', 'e' and 'f' indicate restraints on rotary motion.

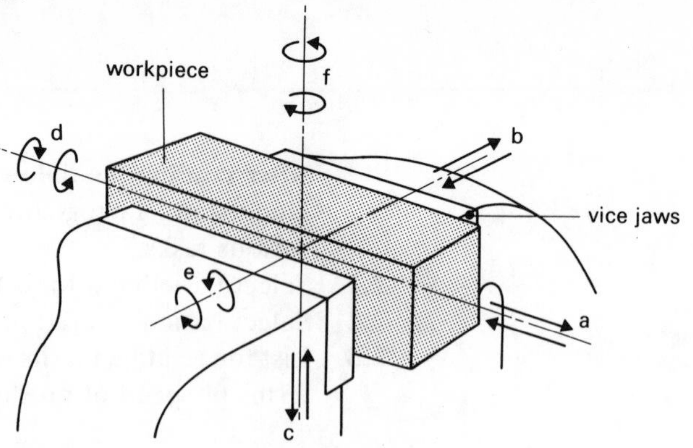

Figure 12 *Restraints on movement of a workpiece in a vice*

In gripping a workpiece, the serrated jaw pieces may damage its surface. If the surface finish of the workpiece is important, vice jaw shoes can be fitted over the serrated surfaces. Figure 13 illustrates a typical assembly.

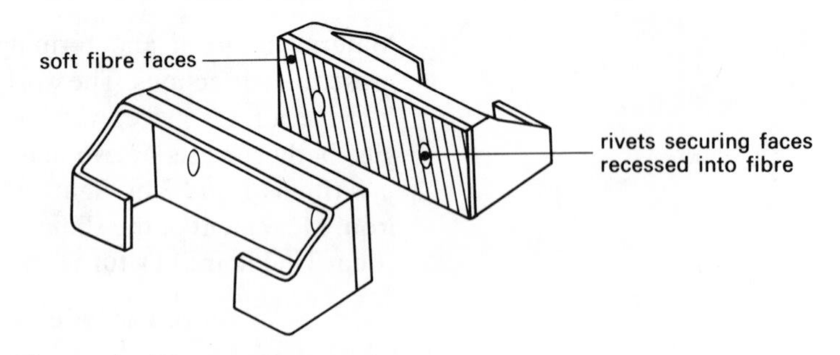

Figure 13 *Vice jaw shoes*

mass of hammer head (kg)	*uses*
0.1–0.25	centre punching and general light work
0.5	light chipping
1.0	chiselling
1.5	heavy work

Figure 14 *Sizes of hammer*

The engineer's hammer

Various sizes of hammer can be obtained; they are classified in terms of mass as shown in Figure 14.

Figure 15 shows the shapes of three hammer heads in general use. Although the striking face remains the same for a particular size of hammer head, the opposite end, called the pein, can take several shapes. For example the ball pein is useful in 'mushrooming' rivet heads and pins. Cross and straight peins can also be used for riveting

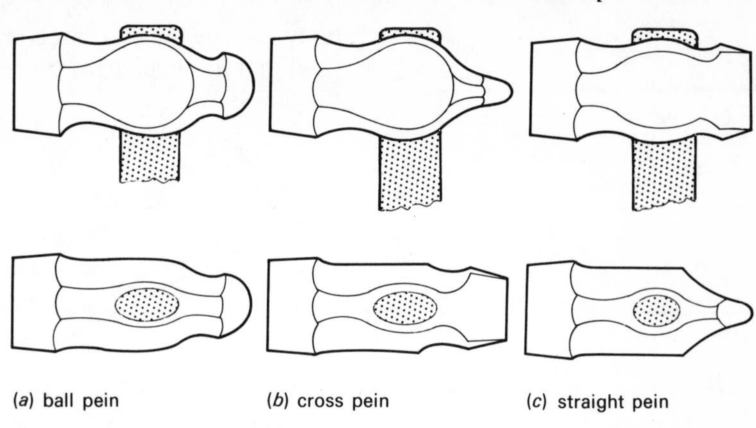

Figure 15 *Fitter's hammers*

when less space is available, but they are useful for peening, an operation used to straighten bent material.

Another type of hammer is used when the work must not be marked; this is the soft-headed hammer (see Figure 16). Such hammers have heads made from rawhide, copper or plastic, set in a cast steel body. Rawhide and copper heads are usually a press fit, but screw fittings are used for plastic heads. Soft-faced hammers should not be used on rough surfaces, and it is important to keep these hammer faces clean and free from metal chips.

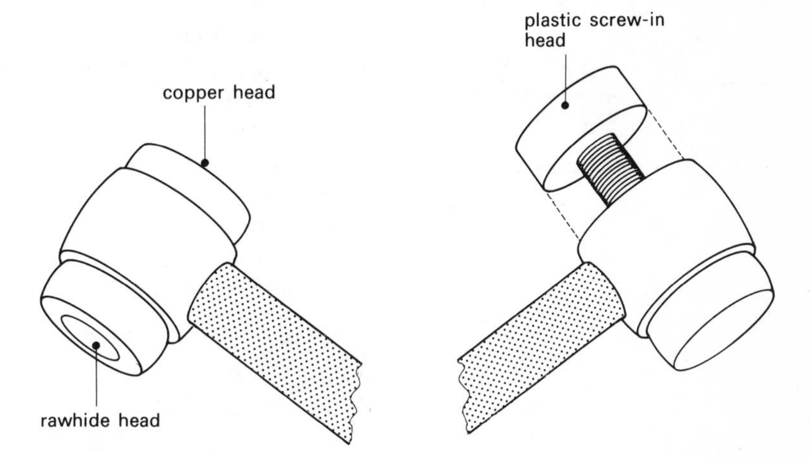

Figure 16 *'Soft' hammers*
Soft-head hammer (press-fit heads)
Soft-head hammer (screw-in heads)

The hand chisel

Chiselling is a method of removing material fairly rapidly. A price is paid for this speed, for the surface finish and geometrical accuracy are usually poor. To be effective, the chisel must be hard but at the same time, because of shock loads applied by the hammer, the body must be tough. Chisels are made from either tool steel or alloy steel that has been heat treated to give the required hardness and toughness.

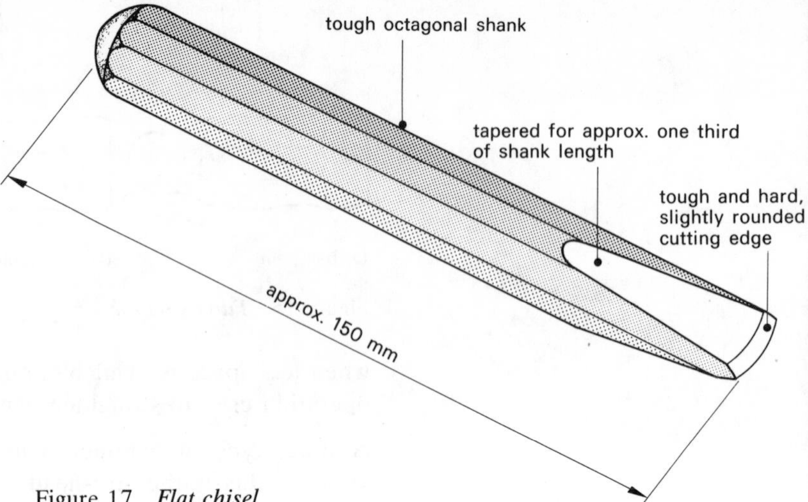

Figure 17 *Flat chisel*

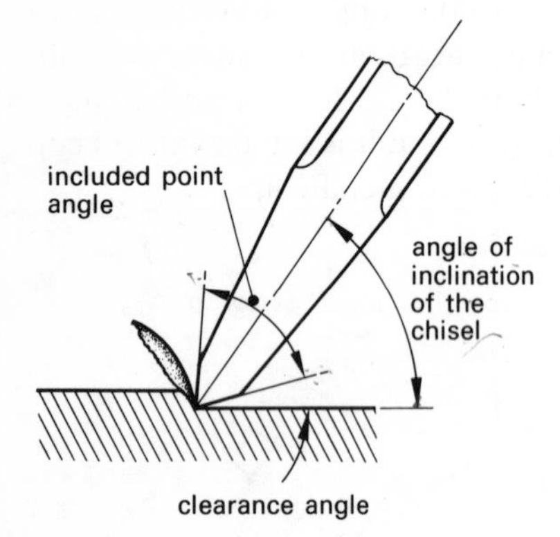

Figure 18 *Cutting angles of a chisel*

The most common chisel is the flat chisel (see Figure 17), and this can be adapted to work on a variety of materials by:
changing the included point angle,
changing the angle at which the chisel is held.
Both of these features are shown in Figure 18.

A cross-cut chisel (see Figure 19) is used for cutting groves. The cutting edge is about 6 mm wide and tapers slightly to a thinner section at the shoulder, so allowing clearance between the chisel and the workpiece. Grooves of smaller section can be cut using the half-round or diamond point chisel as illustrated in Figure 20. Oil grooves in bearing surfaces can be produced by using half-round chisels while vee grooves require the more acute diamond-pointed chisel. One other use of the diamond-pointed chisel is to correct drilled hole centres that have been started in an incorrect position.

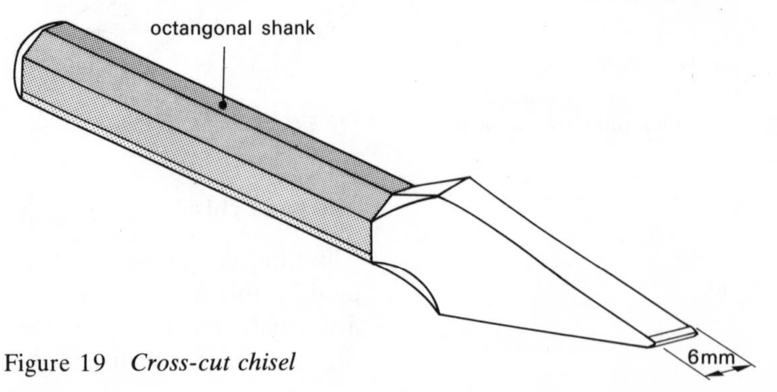

Figure 19 *Cross-cut chisel*

Goggles should be worn at all times when using a chisel in order to reduce the risk of damage to the eyes.

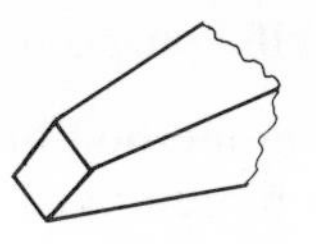

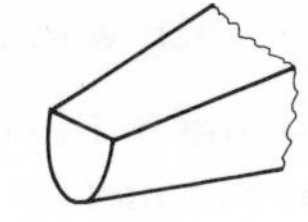

(*a*) diamond point (*b*) half round

Figure 20 *Diamond-pointed and half-round chisels*

The file

Files are supplied in a variety of lengths, shapes and grades of cut in order to suit the wide variety of uses to which they are put. The shape of a file is described by the shape of its cross-section, e.g. square, triangular, round, half-round (the cross-section is not a complete semi-circle, rather it is a segment of a circle), and flat (rectangular in cross-section).

The grade of cut influences the rate of metal removal and depends upon the pitch of the teeth; rough files have teeth set at a larger pitch than smooth files. Also, the pitch of the teeth increases with longer files, and decreases with shorter files. Figure 21 indicates the names given to the grades of file, and suggests typical uses for each grade.

grade	*typical uses*
rough	soft metals
bastard	iron castings and general roughing-out
second cut	roughing-out hard metals, and finishing soft materials
smooth	general finishing and draw-filing
dead smooth	used only when a good surface finish is required

Figure 21 *Grades of file*

Files may be obtained with the teeth cut in either of two patterns:
single-cut teeth which are mainly used on hard materials, or
double-cut teeth which are used for removing softer materials (see Figure 22).

Methods of filing

A file should *never* be used without a handle. The *tang* of a file, although not needle sharp, can cause serious injury to the operator if a file is used without a handle.

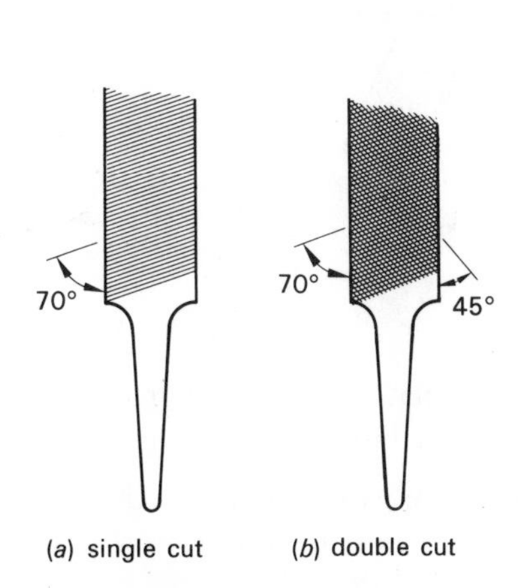

Figure 22 *Single-cut and double-cut teeth*

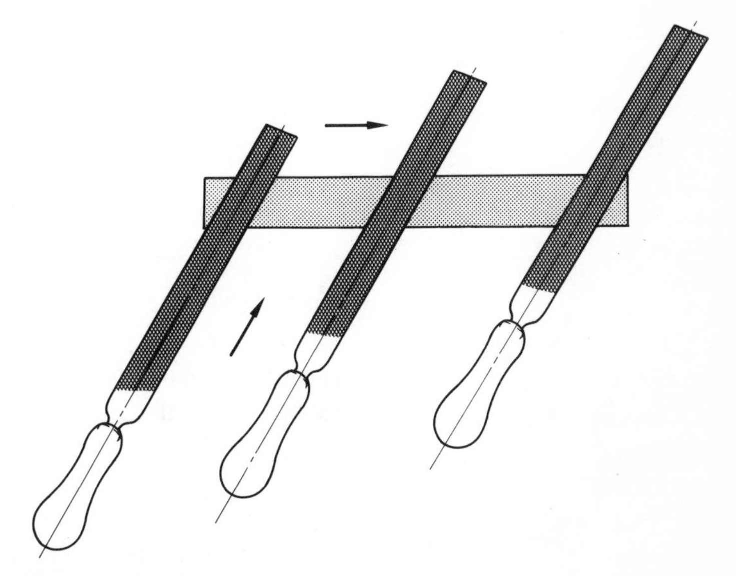

Figure 23 *Cross filing*

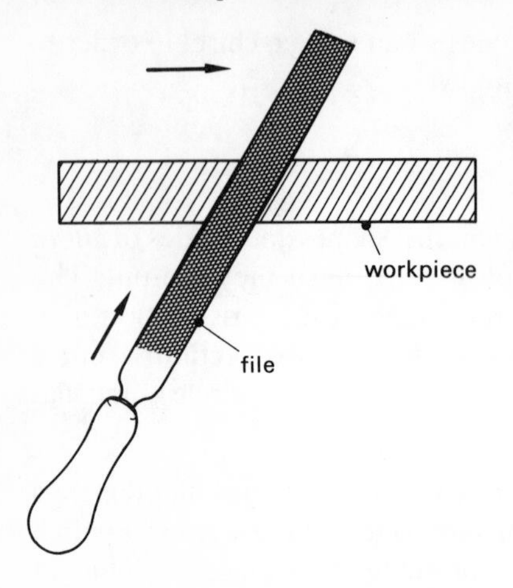

There are several methods of filing, each with a specific purpose.

Cross filing (Figure 23) is probably the most common method of filing; it is used when large amounts of metal are to be removed. Figure 24 illustrates the file being used across one diagonal for a short period of time, and then across the other diagonal. This ensures that the maximum amount of metal is removed, and that the possibility of 'rounding' the surface is reduced.

Straight filing (Figure 25) is used when a short length of workpiece is required to have a flat surface. File marks made during cross filing may be removed to produce a relatively smooth surface.

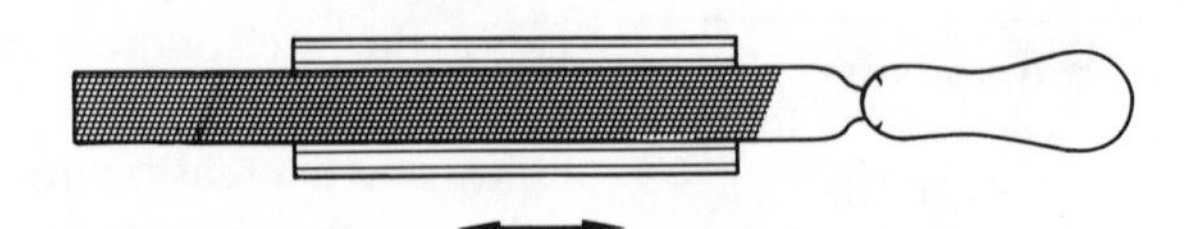

Figure 25 *Straight filing*

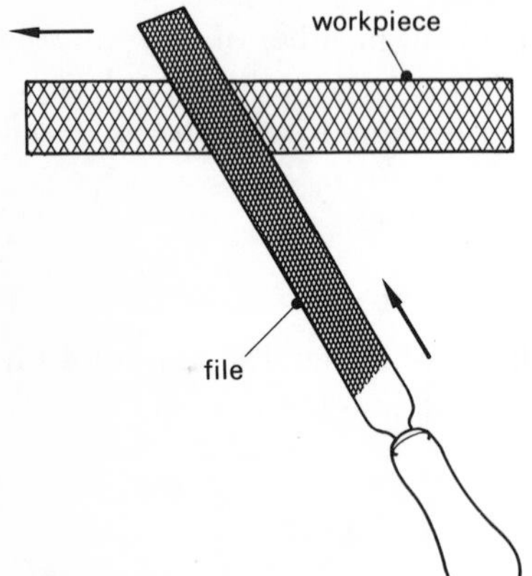
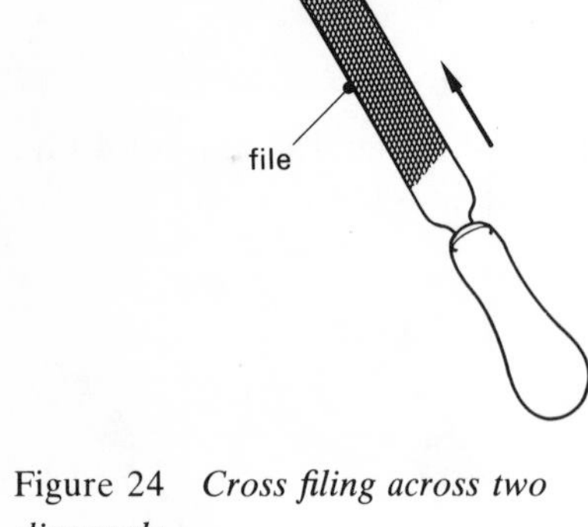

Figure 24 *Cross filing across two diagonals*

Draw filing (Figure 26) produces a smoother surface finish than straight filing. A smooth or dead smooth flat file is used.

During filing operations, material becomes clogged in the teeth, and hence reduces the effectiveness of the file. These particles may be removed by using a file card.

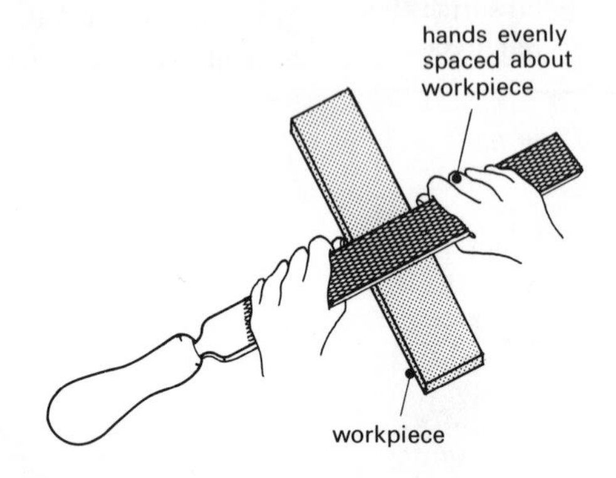

Figure 26 *Draw filing*

steel bristles

wooden base

Figure 27 *File card*

(Figure 27). The file card is a small flat wire brush which, when rubbed across the teeth, dislodges unwanted material.

The teeth on a new file are sharp. To prevent damage to the hard, brittle teeth, the file should be used initially on relatively soft materials such as copper or aluminium. As the file wears, mild steel and harder materials can be worked. Teeth can also be damaged by accidental knocks; after use, a file should always be stored in a rack.

The scraper

In using a file or chisel, relatively large amounts of material can be removed quickly, but the resulting surface is far from smooth. The surface finish can be improved by *scraping*, a 'finishing' process that entails removing small amounts of material. When examined under a microscope, the surface of a filed component has high and low spots very similar to hills and valleys (Figure 28). Scraping reduces the height of the high spots. Scraping is carried out on the slideways of metrology equipment, machine beds, etc.

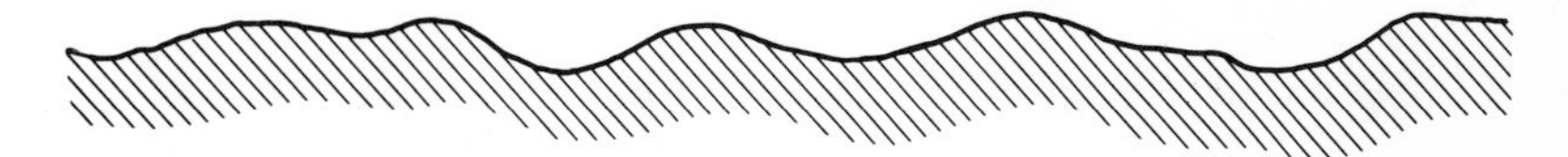

Figure 28 *Surface irregularities on what may appear to be a flat filed surface (magnified)*

Three main types of scrapers are in general use. The flat scraper is the most common and is illustrated is Figure 29. It is used to produce flat surfaces of high quality.

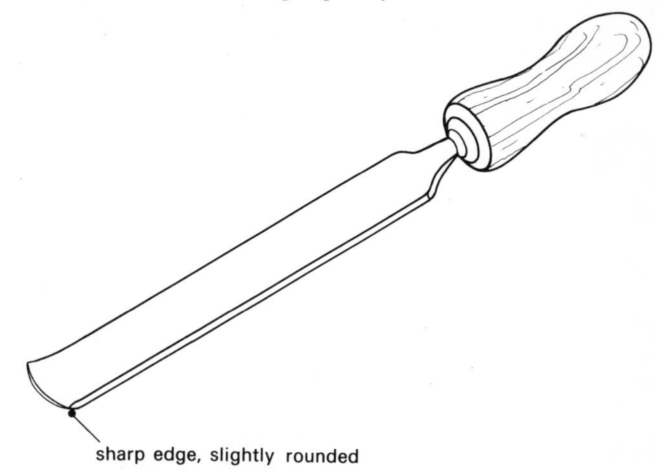

Figure 29 *Flat scraper*

Curved faces are produced by using a half-round scraper (see Figure 30). An example of where a half-round scraper may be used is on white-metal journal bearings. It is also used on bushes, for chamfering holes and for removing burrs.

A third type of scraper (Figure 31) is called a three-square scraper. It is generally used for de-burring holes or other edges that have been left in a sharp, ragged condition.

The cutting edge of a scraper is maintained by sharpening on an oilstone. Because the scraper has a keen edge, it is important to

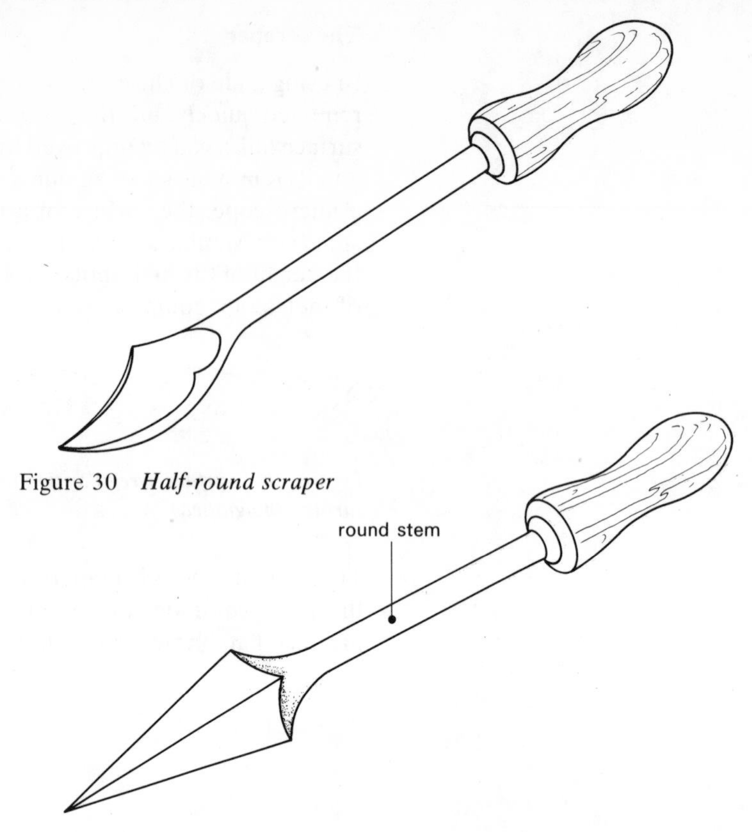

Figure 30 *Half-round scraper*

Figure 31 *Three-square scraper*

store it safely both to protect the edge and to prevent injury to workmen.

The hacksaw

The hacksaw is used for cutting metal, the surface of which can then be finished by an operation such as filing. The hacksaw consists of a steel frame, with a blade which is held by a pin at each end. The blade is tensioned by a wing nut (Figure 32).

Blades are normally stocked in a standard length of approximately 300 mm, but the pitch of the teeth of blades varies. A blade should be chosen so that at least two teeth are always in contact with the work (Figure 33). On thin materials a fine pitch blade should be used, typical teeth pitches being from 1.4 mm to 0.8 mm. When cutting very thin plates, the metal is frequently held firmly between two pieces of wood. This helps to prevent fracture of the teeth of the blade.

Once an appropriate blade has been selected, it should be secured

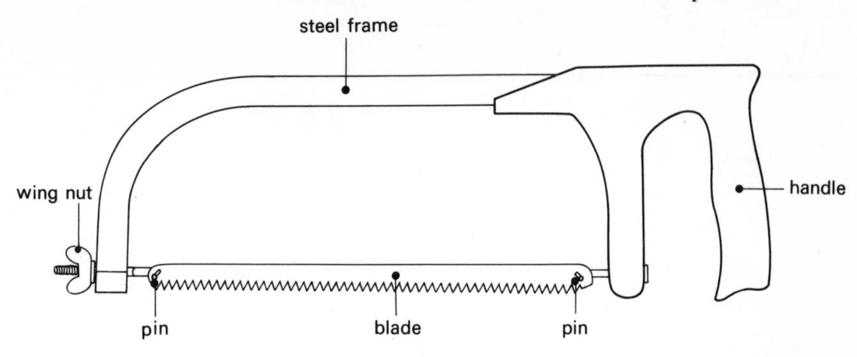

Figure 32 *Hacksaw*

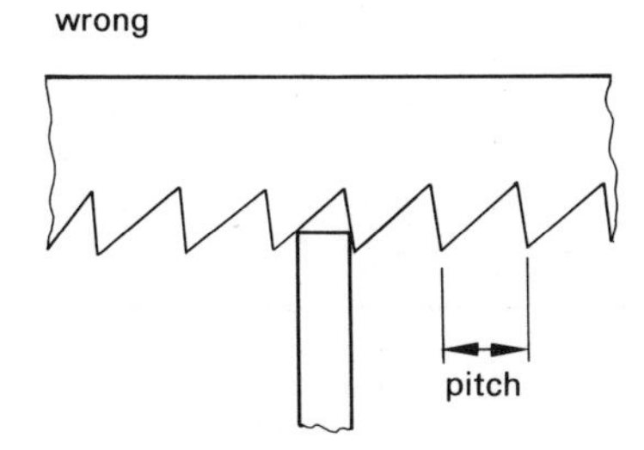

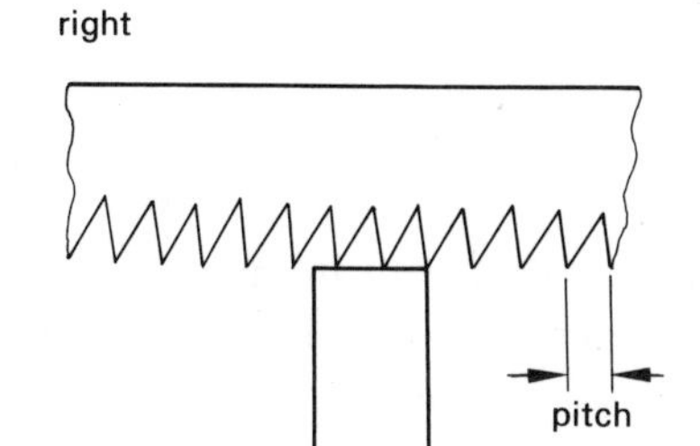

Figure 33 *Correct pitch of hacksaw blade*

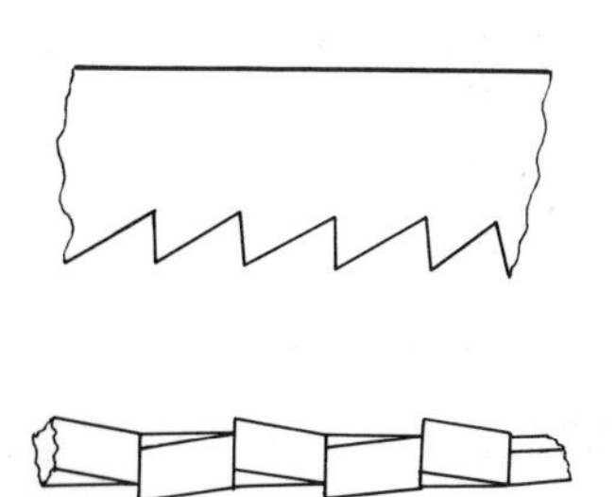

Figure 34 *Allowance for clearance by setting the teeth*

firmly in the frame, with the points of the teeth pointing away from the handle. The cutting stroke is the forward stroke away from the operator; the return stroke is to clear the teeth of sawn metal. This operation is facilitated because the teeth are set at a slight angle to the direction of cut, as shown in Figure 34. The set of the teeth also allows clearance of the blade in its cut; if there is no clearance between the blade and the cut metal, the blade tends to seize.

Powered hand tools

Examples of powered hand tools include the drill, saw, grinder, screwdriver, spanner and riveter. The most usual source of power is electricity, although pneumatic (compressed air) drills, riveters and spanners have many applications.

The basic safety principles outlined in the previous topic area should always be adhered to when using powered hand tools. Goggles should always be worn if metal is being cut, and ear muffs protect the ears when using a powered riveter.

The operator should check that the flexible cable which connects the drill to the supply is mechanically sound. He should also be assured that the drill is safe to operate and is designed to work on the electricity supply to which it is to be connected.

If the source of power is compressed air, airlines and connections should be thoroughly checked for leaks; the tool should be securely fastened to the airline *before* the air valve is opened. Compressed air contains a large amount of energy which can be very destructive if released suddenly.

The cost of a powered hand tool is higher than its manually operated counterpart. However, the speed of operation of the powered tool is greater, and so the initially higher cost may be repaid over a period of time (Figure 35).

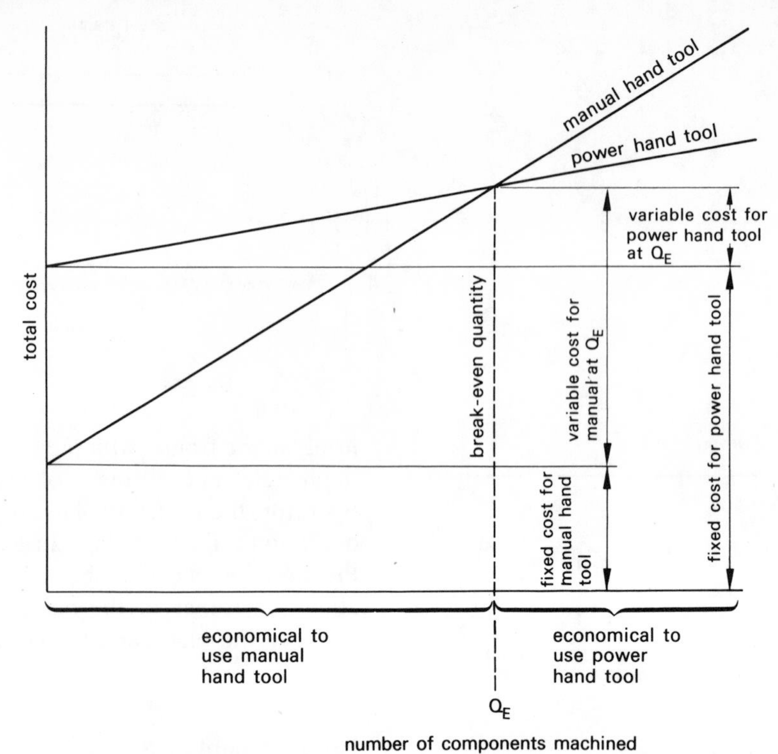

Figure 35 *Relative costs of powered and manually operated hand tools*

In general, a powered hand tool such as a drill produces more accurate machining than a manually operated tool. The reasons for increased accuracy may be summarized as follows:

Operator fatigue is reduced.

The operator has only to concentrate on guiding and supporting the tool; most of the power is supplied by the machine.

There is consistency throughout the operation due to a more appropriate operational speed than can be achieved with manually operated tools.

Stands can often be utilized to clamp the tool, so leaving the operator with both hands free to guide the tool.

Self-assessment questions

1. Explain why the jaws of a fitter's vice are serrated.
2. Describe how the surfaces of a component can be protected when clamped in a fitter's vice.
3. Show by means of a sketch how the movements of a workpiece are restrained when it is clamped in a fitter's vice.

4 For what purpose is a soft-headed hammer used?

5 What mechanical properties are required in
(a) a chisel shank
(b) a chisel cutting edge?

6 Name four common types of chisel used by fitters.

7 For what purpose should a cross-cut chisel be used?

8 Name three grades of hand file, and give examples of their uses.

9 Give examples of where the following hand files are used:
(i) square file
(ii) half-round file
(iii) three-square (triangular cross-section) file

10 State when each of the following methods of filing should be used:
(i) cross filing
(ii) draw filing
(iii) straight filing

11 Describe a file card and explain why it is used.

12 When is a scraper used?

13 Which way should the teeth of a hacksaw blade be pointing when fitted into the frame of a hacksaw?

14 Give an example of when a hacksaw blade with
(i) a fine tooth pitch and
(ii) a coarser tooth pitch
should be used.

15 Name two sources of power most commonly used to operate powered hand tools.

16 Give three examples of powered hand tools normally found in an engineering workshop.

17 List the advantages of powered hand tools when compared with manually operated hand tools.

18 Describe some of the safety precautions which should be observed when using electric powered hand tools.

After reading the following materials, the reader shall:

3 Be aware of the need for marking out of components.
3.1 Describe the function of datum lines and centre lines.
3.2 Select basic marking out equipment to perform simple tasks, such as use of dividers, rules, vee blocks, surface gauge to mark out profiles.
3.3 Describe the use of a marking-out table.

If relatively small quantities of a component are to be machined, it is usually uneconomic to use automatic machine tools or special jigs and fixtures. A more accepted method is to mark the main features of the finished component on to the surface of the material to be machined. This method is known as *marking out*. Sometimes, when producing large quantities of a component using jigs and fixtures, it may be necessary to carry out some initial marking out and machining. Other characteristics of marking out include:

(i) A guide is given as to the amount of material to be removed from the workpiece.
(ii) The position of holes, slots, keyways and other features is indicated.

In order to mark out accurately, all measurements must be taken from a fixed point, a line or a face. These are referred to as datums. Taking individual measurements from the same datum for each feature of the workpiece avoids a cumulative error that otherwise might occur. In marking out flat plates datum faces are used (Figure

Solutions to self-assessment questions

1 The serrations on the jaws of a vice ensure that a component is clamped firmly between the jaws.

2 Protection is given to work clamped in a fitter's vice by placing vice jaw shoes, made from soft material, over the cast steel jaws.

3 Figure 12 (page 22) illustrates the restraints on a workpiece when clamped in a fitter's vice.

4 A soft-headed hammer is used on work whose surface is to have a minimum of marks or scars.

5 The chisel shank should be tough enough to withstand the shock forces applied by hammer blows. The chisel cutting edge should be hard in order to resist deformation whilst cutting; also, it needs to be tough so that it does not chip.

6 Four common types of chisel are a flat chisel, a cross-cut chisel, a diamond point chisel and a half-round chisel.

7 A cross-cut chisel is used for cutting groves that are larger than about 6 mm across.

8 The reader could select from the following five grades of file:
rough – used on soft materials such as aluminium, copper, or brass
bastard – used on cast iron, and when relatively large amounts of materials are to be removed
second cut – used for finishing cuts on soft materials and 'roughing' cuts on harder materials
smooth – used for finishing cuts and draw filing
dead smooth – used when a good surface finish is required

9 (i) Uses of a square file include work on internal square corners, square apertures and slots.
(ii) Uses of a half-round file include forming large internal radii and arcs.
(iii) Uses of a three-square file include work on internal square corners and apertures, and internal corners containing acute (less than 90°) angles.

36). Note that the datum faces A and B are at right angles to one another.

The method used for positioning holes prior to machining is also shown in Figure 36, making use of two intersecting centre lines. The dimensions referring to the centre lines are known as co-ordinates, and locate the positions of the centres of holes which are to be machined.

The marking out of many components requires the use of three datum faces, as shown in Figure 37. Here the shape of the main features of the components have been defined by using the base (datum face A), one side (datum face B) and an end (datum face C).

To mark out accurately, a datum surface of known accuracy is used. Such a surface is frequently provided by a marking-out table. The degree of 'flatness' of the surface of a marking-out table is defined within strict limits – much smaller limits than can be achieved using

10 (i) Cross filing is normally used for the quick removal of material.
(ii) Draw filing is used to produce a good surface finish.
(iii) Straight filing is normally used to produce a flat surface on a component which is relatively short in length.

11 A file card is a metal bristled brush that is used to remove small particles of material from the teeth of a file.

12 A scraper is used when a surface of high geometrical accuracy is required.

13 A hacksaw blade should be fitted so that the points of the teeth are pointing away from the handle of the frame of the hacksaw.

14 A hacksaw blade should be chosen so that at least two teeth are always in contact with the work being cut. Thus, thin sheet material requires a fine pitched blade, but thicker material can be cut with a coarser pitched blade.

15 The most common sources of power for powered hand tools are electricity and compressed air.

16 Typical examples of powered hand tools include drills, grinders, screwdrivers, riveters and saws.

17 Compared with manually operated tools, the advantages of powered hand tools include:
(i) cheaper production costs over long periods of time, or when large numbers of components are involved
(ii) consistent cutting speed
(iii) freedom of hands for guiding the tool
(iv) less physical effort required

18 When using powered hand tools to cut metal, goggles should always be worn. Electric tools must only be used with the correct voltage; they should be correctly earthed; the circuit should contain a fuse; the insulation around the cable should be undamaged and the cable should be of adequate length.

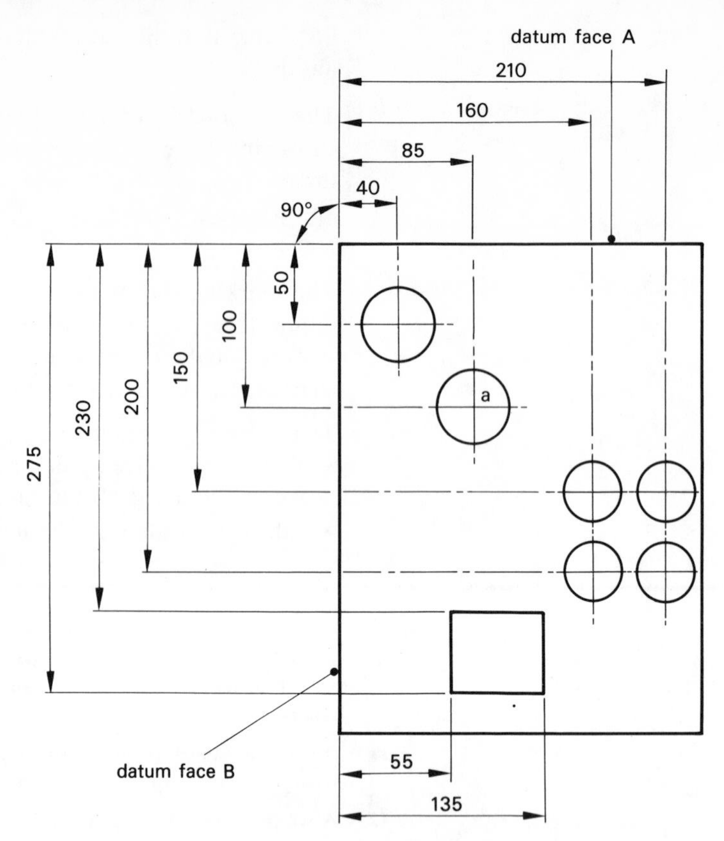

Figure 36 *Locating features using two datum faces. All dimensions in millimetres*

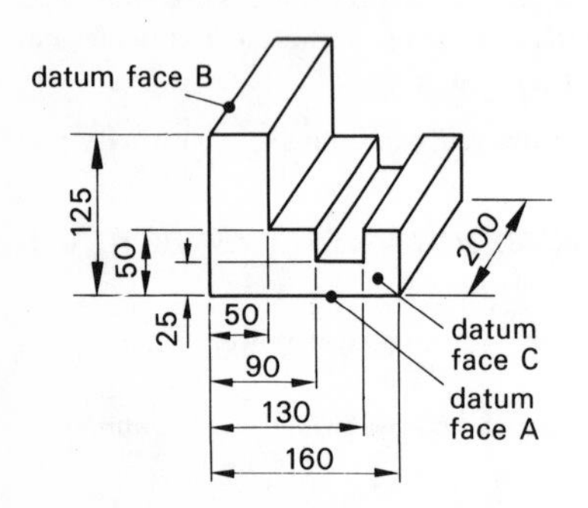

Figure 37 *Locating features using three datum faces. All dimensions in millimetres*

standard machining techniques. The surface is made extremely hard, and therefore resists scratches and indentations.

Several items of equipment are used in marking out operations. One of the most versatile is a steel rule. As the steel rule is used for line measurement, care should be taken to protect its ends as these provide the datums from which measurements are taken. Two examples of taking a measurement using a steel rule are shown in Figure 38. Figure 38 (a) shows the datum for measurement being formed by placing the end of the steel rule against part of a component, whilst in Figure 38 (b) the datum is provided by the flat surface of a marking-out table. With care, a steel rule can be used to measure to an accuracy of about 0.25 mm.

To mark out circular features use is made of dividers (see Figure 39). Co-ordinates are marked on a workpiece, their intersection being clearly defined by marking with a centre punch. This centre punch mark is used to position one leg of the pre-set dividers, whilst the other leg is swung around to scribe the required line.

The scribing block is used to mark lines that are parallel to a datum. An example is shown in Figure 40.

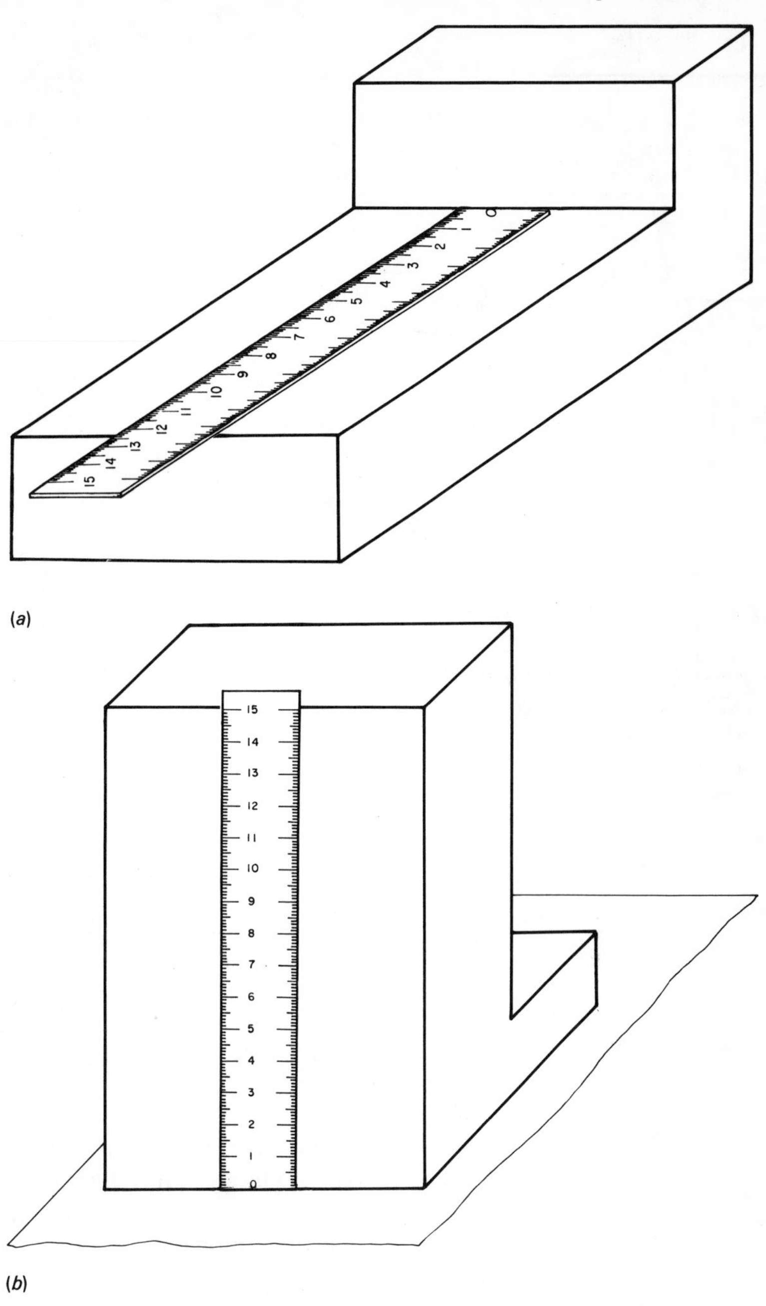

Figure 38 *The engineer's steel rule: using one end as a datum from which to measure*

To set a scribing block to a particular dimension, use can be made of a steel rule positioned at 90° to the datum (Figure 40).

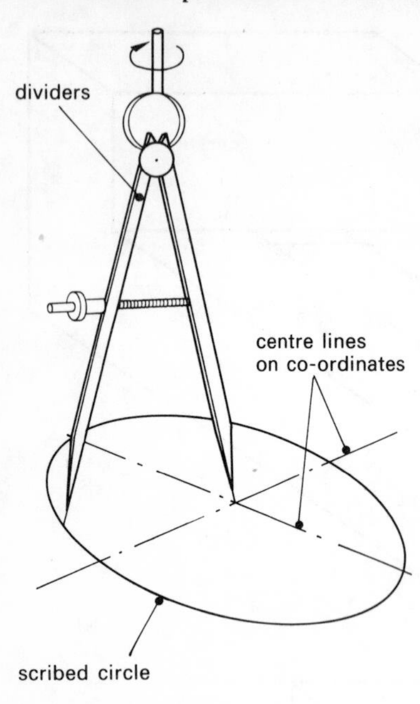

Figure 39 *Dividers being used to scribe a circle on to a workpiece*

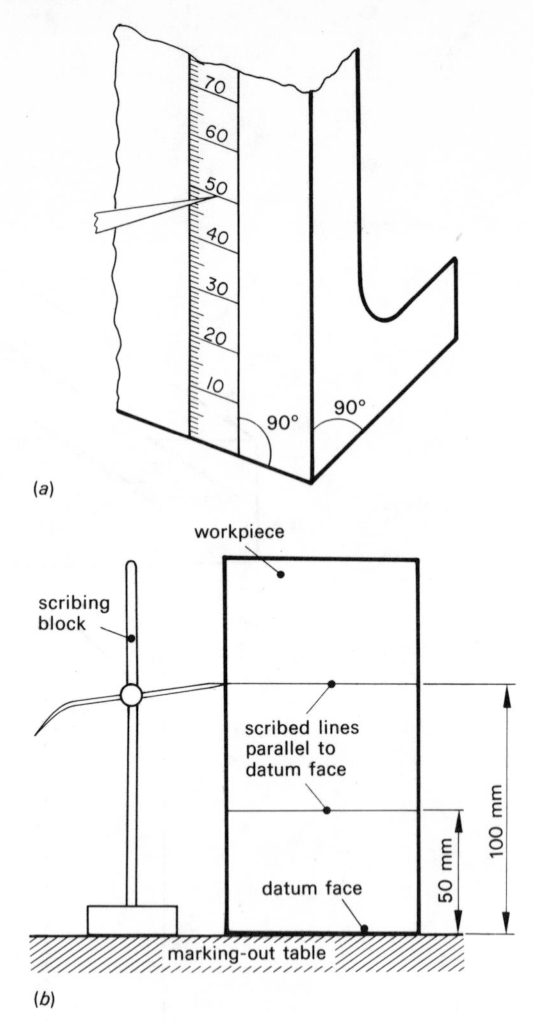

Figure 40 *Using a scribing block*

Figure 41(a) shows the main items that go to make up a scribing block. On an adjustable post scribing block (Figure 41(b)), fine adjustments to the height of the scriber point from a datum can be made by tilting the post on a rocker arm mechanism. A fine adjustment screw is provided at one end of the rocker arm.

The marking out of round components requires the use of vee blocks (Figure 42). It is important that the vee block used is large enough to accommodate the workpiece.

Figure 43(a) shows a workpiece correctly located in a vee block; Figure 43(b) shows a workpiece too large for the vee blocks being used.

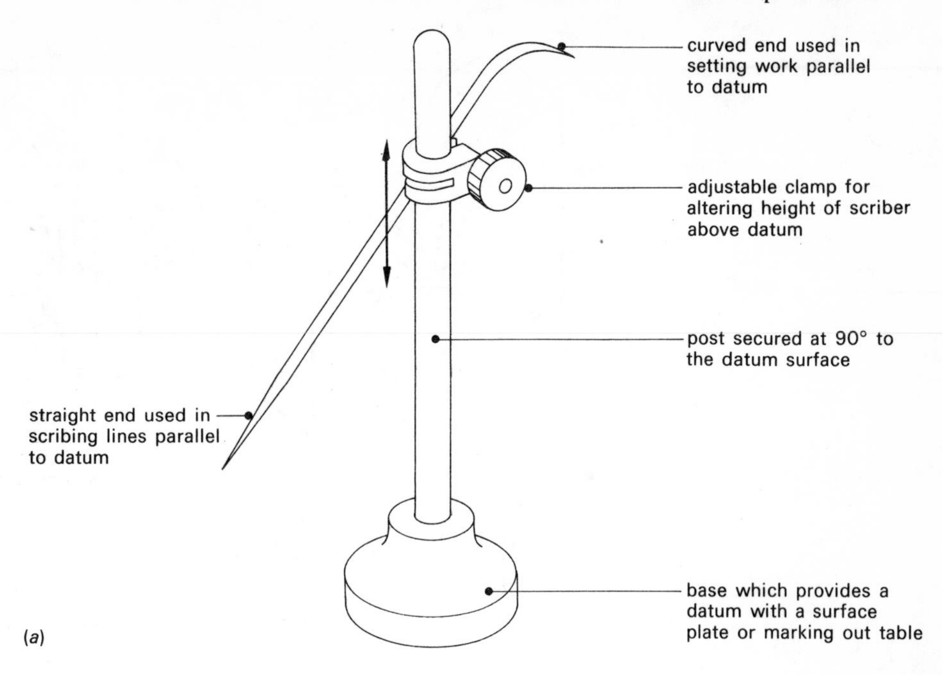

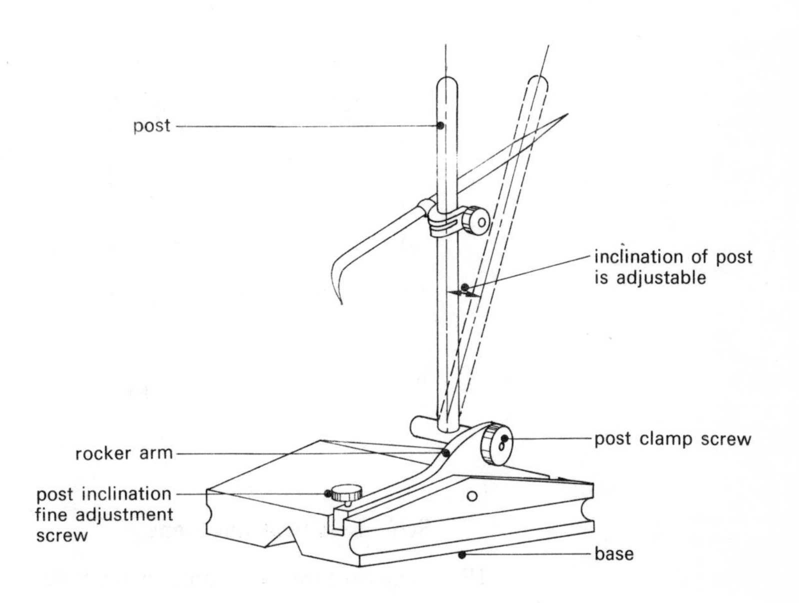

Figure 41 *Scribing blocks*

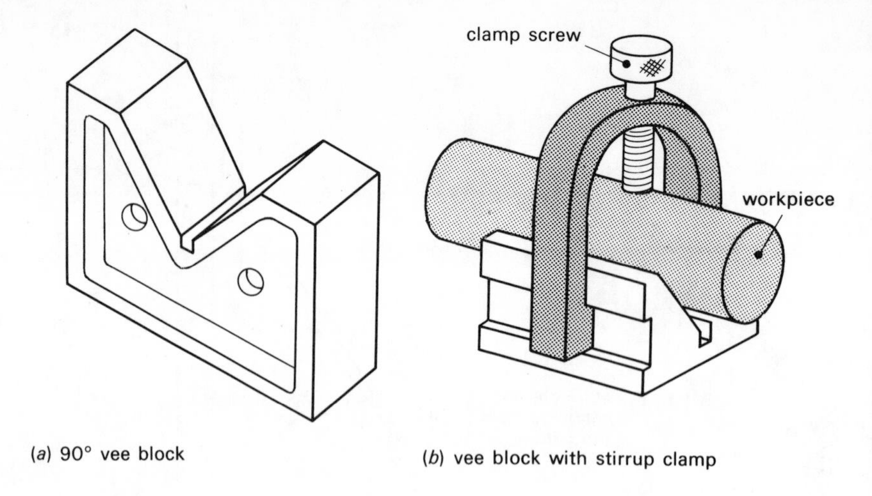

Figure 42 *Use of vee blocks*

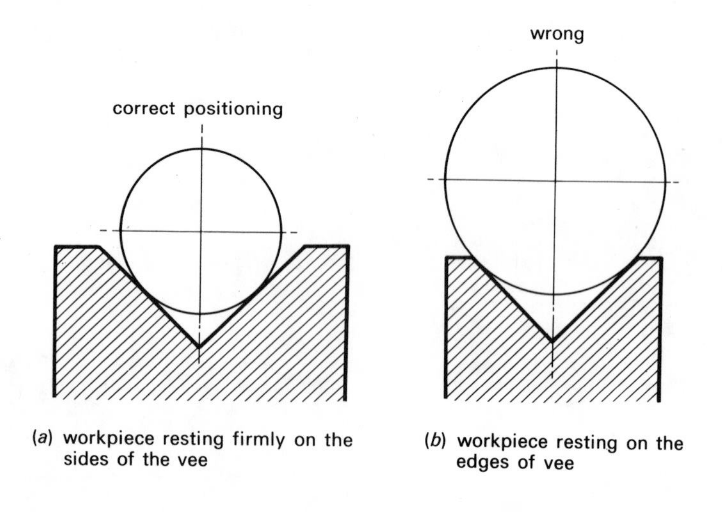

Figure 43 *Using a vee block correctly*

Self-assessment questions

19 What are the co-ordinates for hole 'a' in Figure 36 (page 34) from datum faces A and B respectively?

20 Explain what is meant by a datum.

21 Figure 44 shows a finished component that is machined from a casting. Sketch the component and re-dimension it, indicating clearly the datums used.

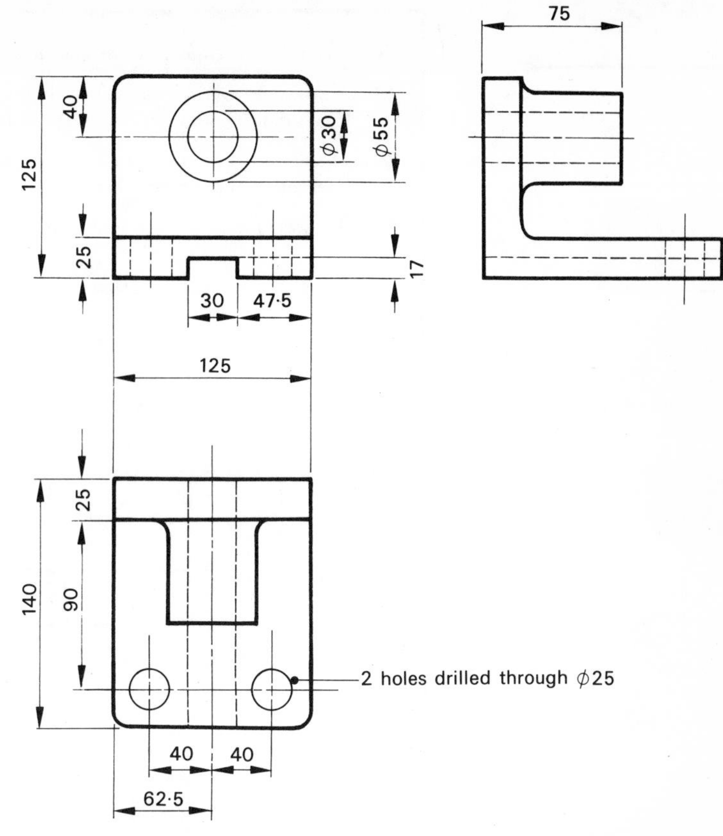

Figure 44 *Self-assessment question 21. First-angle projection. All dimensions in millimetres*

22 Explain how the position of a hole is located when marking out a component.

23 Describe the function of a marking-out table.

24 Describe the main function of a scribing block.

25 Explain with the aid of a sketch the use of vee blocks in marking-out operations.

After reading the following material, the reader shall:

4 Understand the principle and use of simple measuring equipment.
4.1 Describe the principle of the non-digital micrometer.
4.2 Read the scales of external, internal and depth micrometers.
4.3 Describe the principle of the vernier.
4.4 Read the scales of caliper, height, depth and protractors with vernier scales.

Solutions to self-assessment questions

19 Co-ordinate of hole 'a' from datum face A is 100 mm.
Co-ordinate of hole 'a' from datum face B is 85 mm.

20 A datum is a reference in the form of a fixed point, a line or a surface, from which measurements can be taken.

21 Figure 45 indicates the three datums used to re-dimension the component.

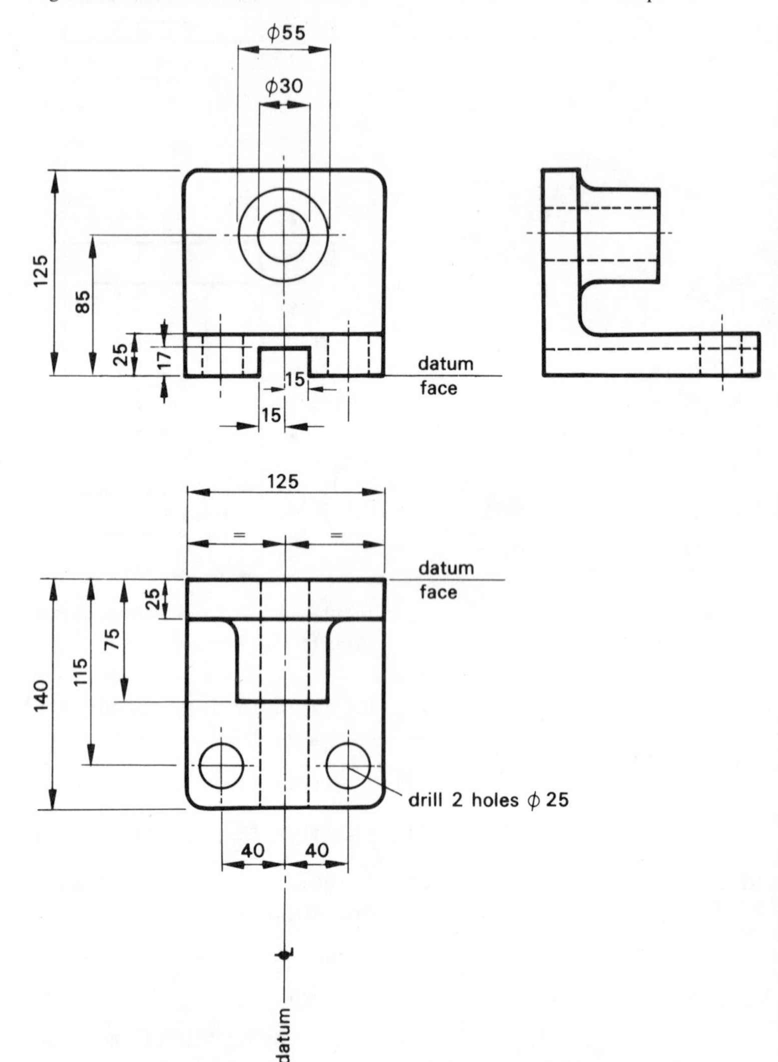

Figure 45 *Solution to self-assessment question 21. First-angle projection. All dimensions in millimetres*

22 The centre position of a hole is marked by centre lines. These are located by co-ordinates, each of which is measured from a datum.

4.5 Identify the different applications of plunger dial gauges and lever type test indicators.

4.6 Identify the limitations of micrometers, verniers and dial gauges from the point of view of accuracy, robustness, etc.

Since civilization began, mankind has used many different standards of measurements – particularly in measuring length. Many of the early standards were based upon the length of some part of a human limb; there are many biblical references to a 'cubit', the distance between elbow and finger tips; an English king defined a 'yard' (22 of which constitute a cricket pitch) as the distance from the tip of his nose to the tip of his outstretched fingers.

These standards of length measurement are inappropriate to modern society. Indeed, it was the industrial revolution which gave the main impetus not only to defining standards, but also to methods of measurement. As early as 1805, Henry Maudslay had made a micrometer screw. Disputes had occurred in Maudslay's workshop over the accuracy of measurements carried out by his workmen. To settle such disputes Maudslay used his micrometer, which became known as the 'Lord Chancellor'.

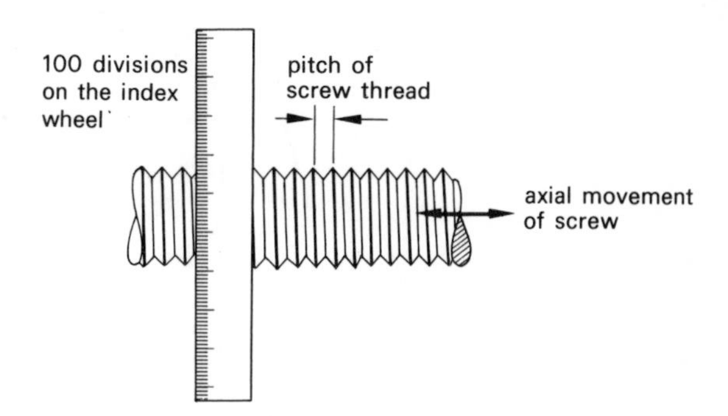

Figure 46 *Basis of Henry Maudslay's measuring machine*

Solutions to self-assessment questions

23 A marking-out table provides a flat plane upon which a component is placed. The surface of the marking-out table is then used as a datum from which measurements are taken.

24 A scribing block makes use of a datum surface (e.g. a marking-out table) to scribe on to workpieces lines that are parallel to that datum surface.

25 Vee blocks are used in conjunction with a marking-out table. They are used to support for example a cylindrical component so that its axis is parallel to the surface of the marking-out table. Typical arrangements are shown in Figures 42 and 43.

This early micrometer consisted of a screw which was coupled to an index wheel graduated into 100 divisions (Figure 46). One turn of this wheel was equivalent to an axial movement of one pitch of the screw. As the wheel was divided into 100 divisions, each division represented $\frac{1}{100}$ of the pitch of the screw. The principle of Maudslay's micrometer remains the basis of modern micrometers.

The micrometer

On a modern micrometer the pitch of the screw thread is 0.5 mm, which means that one revolution of the screw moves it axially 0.5 mm. The screw is fixed to a graduated collar called the thimble (item 2 in Figure 47), which has 50 equal divisions on its circumference:

$$\therefore \text{one division on the thimble} = \frac{0.5}{50}\text{ mm}$$
$$= 0.01 \text{ mm}$$

The basic components of the micrometer screw, that is the spindle (item 1 in Figure 47) and the thimble (item 2 in Figure 47) can be used in a variety of ways.

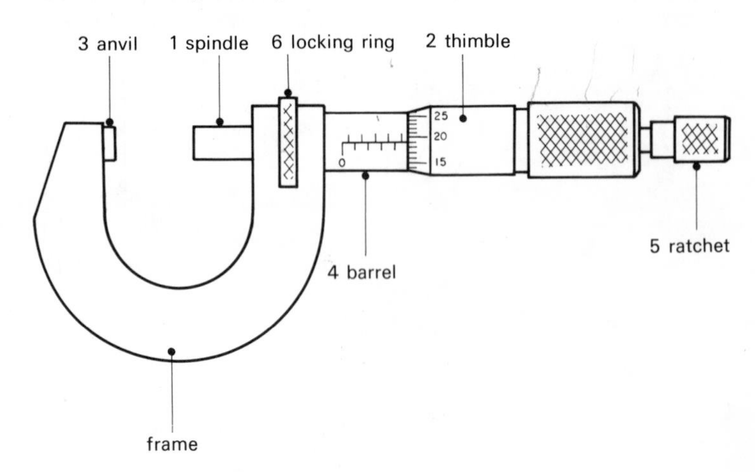

Figure 47 *Outside micrometer*

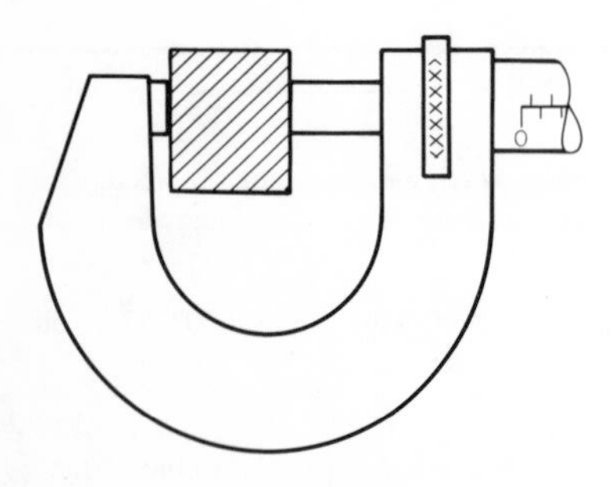

Figure 48 *An outside micrometer being used to measure across flats*

Before a micrometer is used to measure the size of a component it is necessary to zero the instrument. To do this the thimble is rotated until the two anvils are touching and the ratchet slips. At this point a reading is taken; this should read zero, i.e. the thimble should coincide with the zero reading on the barrel and the zero marking on the thimble coincide with the barrel datum marking. If the reading is not zero then the discrepancy should be allowed for in all further readings.

Figure 48 shows an outside micrometer being used to measure

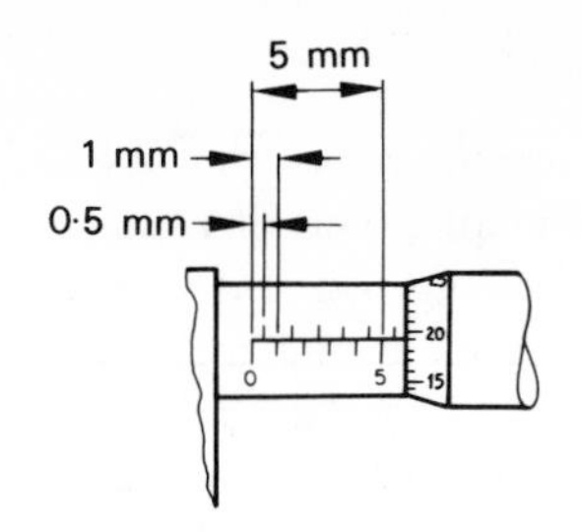

Figure 49 *Micrometer reading*

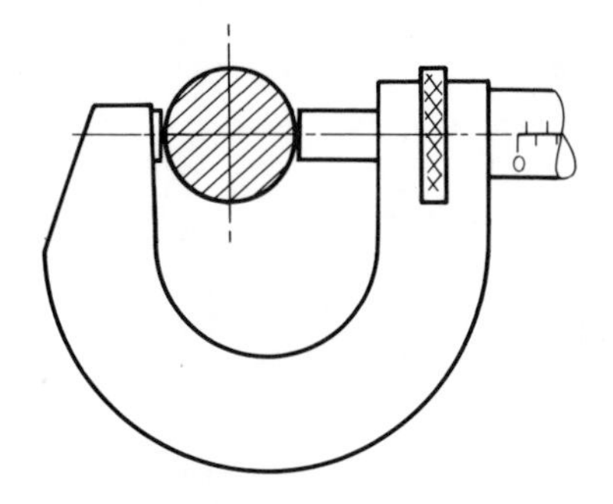

Figure 50 *Using an outside micrometer to measure a diameter*

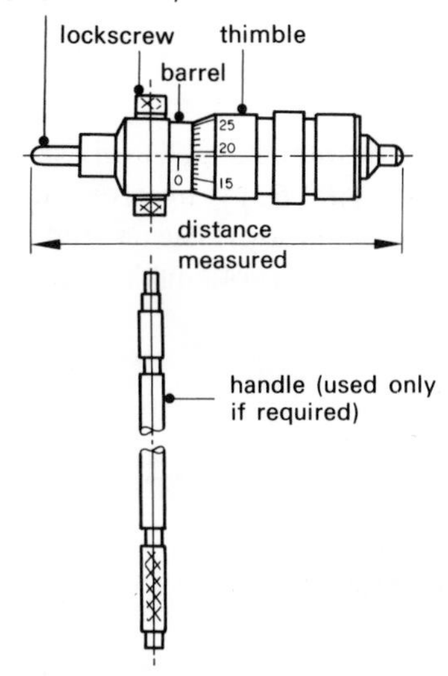

Figure 51 *Internal micrometer*

across the faces of a component; the component is placed between the anvil (item 3 in Figure 47) and the spindle (Figure 48).

The thimble, which is locked to the spindle, is turned until contact is just made with the component. A measurement is taken by noting the position of the thimble in terms of the graduations around its circumference and its position on the barrel (item 4 in Figure 47). A consistent degree of pressure or 'feel' is applied by the micrometer spindle on the component by a ratchet (item 5 in Figure 47), which is fitted to the end of the thimble. At a set pressure, the ratchet slips. At this setting the locking ring (item 6 in Figure 47) is tightened, and the reading taken. To take the reading, the number of divisions visible on the main barrel scale is first noted; every two divisions represent a spindle movement of 1 mm. Then the graduation on the circumference of the thimble that is adjacent to the barrel markings is noted. This reading is now added to the reading noted on the barrel.

For example, Figure 49 shows that eleven divisions on the barrel are visible; this means the spindle has moved 5.5 mm from its zero position. On the thimble the nineteenth graduation is adjacent to the barrel marking, thus:

$$\begin{aligned}\text{the reading is } & 5.5 + (19 \times 0.01) \text{ mm}\\ &= 5.5 + 0.19 \text{ mm}\\ &= 5.69 \text{ mm}\end{aligned}$$

Outside micrometers are manufactured in a variety of sizes in steps of 25 mm: 0–25 mm, 25–50 mm, etc. They are used not only for measuring across flat surfaces but also for measuring diameters.

When using a micrometer to measure a diameter, it is important that the anvil and spindle be situated on an axis passing through the centre point of the component as shown in Figure 50.

The internal micrometer

Figure 51 shows an internal micrometer used for taking internal measurements which are found in cylinder bores, bearing diameters, etc. The measuring faces are situated at each end of the micrometer, and are radiused to give clearance when measuring curved surfaces. Measurements are read in exactly the same manner as with an external micrometer, but unlike the external micrometer, the internal micrometer can be extended to measure a wide measuring range by using spindle extension rods. For example, a basic internal micrometer is capable of measuring diameters between 50–200 mm.

The depth micrometer

The depth micrometer (Figure 52) is another measuring instrument which uses the micrometer screw principle. The measurement is made between the end face of a measuring rod and a measuring face. Because the measurement increases as the measuring rod extends from the face, the readings on the barrel are reversed from the normal: they start at a maximum (when the measuring rod is fully extended from the measuring face), and finish at zero (when the end of the measuring rod is flush with the face).

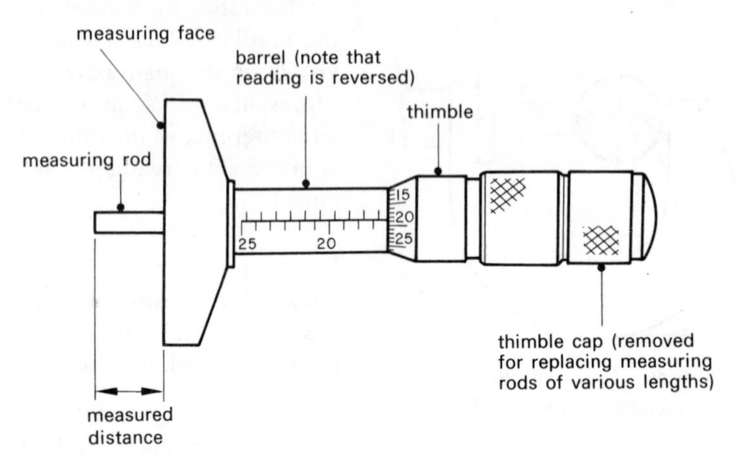

Figure 52 *Depth micrometer*

For example, the measurement on the depth micrometer shown in Figure 52 is:

$$16 + (19 \times 0.01) \text{ mm}$$
$$= 16 + 0.19 \text{ mm}$$
$$= 16.19 \text{ mm}$$

Measuring rods in steps of 25 mm can be interchanged to give a wide measuring range. The thimble cap (Figure 52) is unscrewed from the thimble which allows the rod to be withdrawn. The desired rod is then inserted and the thimble cap replaced, so holding the rod firmly against a rigid face (Figure 53).

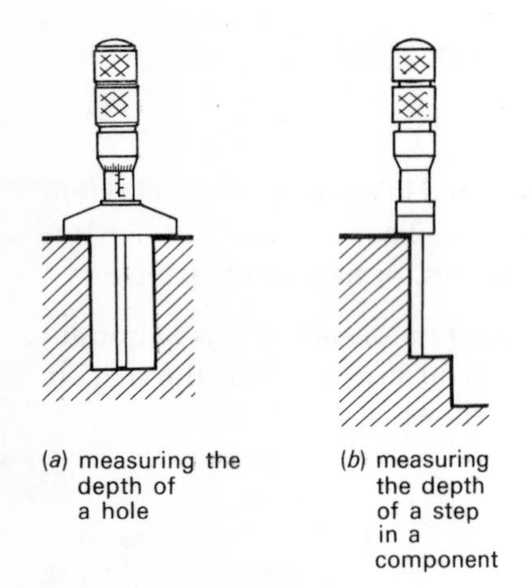

Figure 53 *Using a depth micrometer*

Self-assessment questions

26 Figure 54 shows three micrometer readings; what are the measurements they represent?

27 Figure 55 shows three depth micrometer readings; what are the measurements they represent?

28 Describe the principle of a micrometer screw measuring device, using sketches where appropriate.

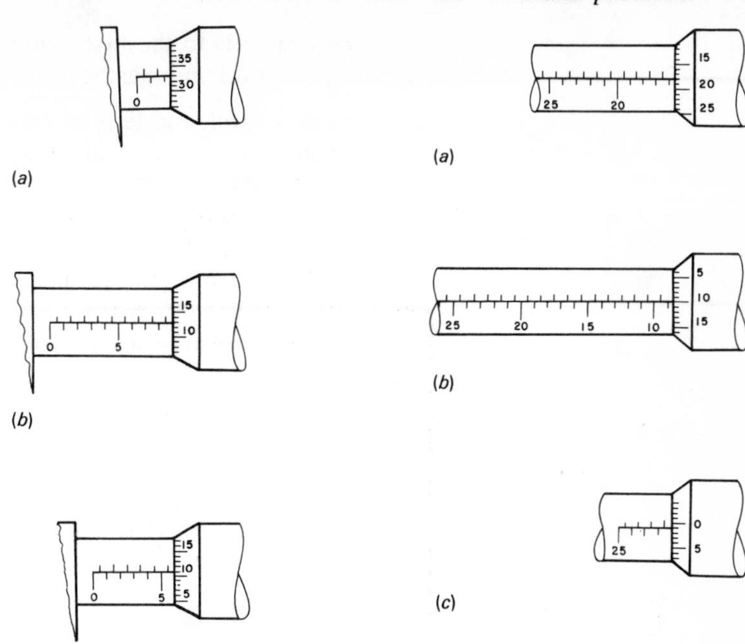

(c)

Figure 54 *Self-assessment question 26*

Figure 55 *Self-assessment question 27*

29 State a particular precaution which must be taken when using an external micrometer to measure a diameter.

30 Why is it important to use the ratchet when taking measurements with a micrometer?

The vernier scale

The micrometer makes use of an accurate screw thread to enable end measurements to be made. The vernier employs a line scale to measure end measurements. The original instrument dates back to Pierre Vernier, who patented the idea in 1631. The principle of the vernier scale is shown in Figure 56(a) where a scale that is marked in divisions of 1 mm is illustrated.

Referring to Figure 56(a), the main scale is accurately graduated in 1 mm steps, and terminates in the form of a caliper jaw. There is a second scale which is moveable, and is also fixed to a caliper jaw. The moveable scale is equally divided into 10 parts but its length is only 9 mm; therefore one division on this scale is equivalent to $\frac{9}{10} = 0.9$ mm. This means the difference between one graduation on the main scale and one graduation on the sliding or vernier scale is $1.0 - 0.9 = 0.1$ mm. Hence if the vernier caliper is initially closed, and then opened so that the first graduation on the sliding scale corresponds to the first graduation on the main scale, a distance equal to 0.1 mm has been moved as shown in Figure 56(b).

Such a vernier scale is of limited use because measurements of a greater degree of accuracy are normally required in engineering workshops. The vernier scale shown in Figure 57 has main scale graduations of 0.5 mm, whilst the vernier scale has 25 graduations equally spaced over 24 main scale graduations, or 12 mm. Hence, each division on the vernier scale $= \frac{12}{25} = 0.48$ mm.

Solutions to self-assessment questions

26 Figure 54(a) shows a reading of
$2 + (33 \times 0.01)$ mm
$= 2 + 0.33$ mm
$= 2.33$ mm

Figure 54(b) shows a reading of
$8.5 + (13 \times 0.01)$ mm
$= 8.5 + 0.13$ mm
$= 8.63$ mm

Figure 54(c) shows a reading of
$5.5 + (11 \times 0.01)$ mm
$= 5.5 + 0.11$ mm
$= 5.61$ mm

27 Figure 55(a) shows a reading of
$15 + 0.5 + (18 \times 0.01)$ mm
$= 15 + 0.5 + 0.18$ mm
$= 15.68$ mm

Figure 55(b) shows a reading of
$8 + 0.5 + (10 \times 0.01)$ mm
$= 8 + 0.5 + 0.1$ mm
$= 8.6$ mm

Figure 55(c) shows a reading of
$21 + (1 \times 0.01)$ mm
$= 21 + 0.01$ mm
$= 21.01$ mm

28 The micrometer spindle is fixed to a screw thread having a pitch of 0.5 mm, so that one revolution of the spindle moves it axially 0.5 mm. Attached to the spindle is the thimble, which is graduated around its circumference into 50 equal divisions. One division is therefore equal to a spindle movement of $\frac{0.5}{50} = 0.01$ mm. Taking a combined reading of the visible barrel markings, which indicate movements of the spindle in steps of 0.5 mm, and the coincident thimble reading gives the measurement required. See Figures 47 (page 42) and 49 (page 43).

29 Care must be taken to position the faces of both the anvil and spindle on an axis which passes through the centre point of the component. In other positions, either above or below the central axis, a smaller measurement than the diameter will be taken.

30 As no two people have the same 'feel' when using a micrometer, different amounts of pressure could be applied between the workpiece and the anvil and spindle. To reduce this inaccuracy a ratchet mechanism enables a consistent pressure to be applied to the faces between which the distance is being measured.

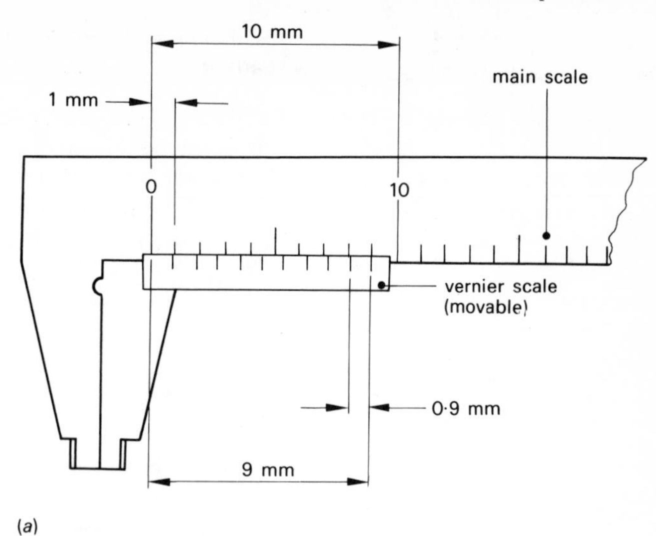

(*a*)

(*b*)

Figure 56
(a) *Principle of a 0.1 mm vernier*
(b) *A 0.1 mm vernier showing a reading of 0.1 mm*

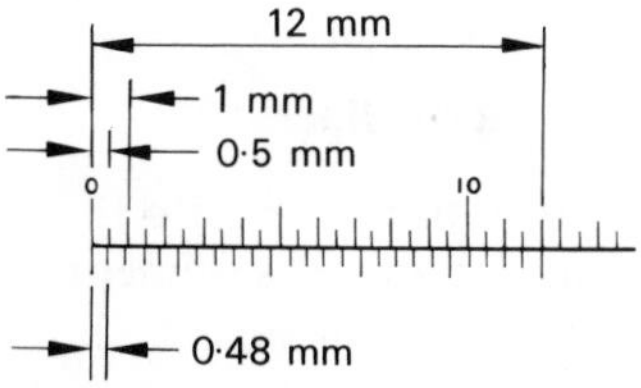

Figure 57 *Principle of a 0.02 mm vernier scale*

The difference between one division on the main scale and one division on the vernier scale is 0.5 − 0.48 = 0.02 mm.

To read this type of vernier, the number of millimetres and half millimetres on the main scale that are coincident with the zero on the vernier scale is noted. Next, the graduation on the vernier scale that coincides with a graduation on the main scale is found. (Close examination is necessary at this stage.) This figure must be multiplied by 0.02 to give the reading in millimetres. The total measure-

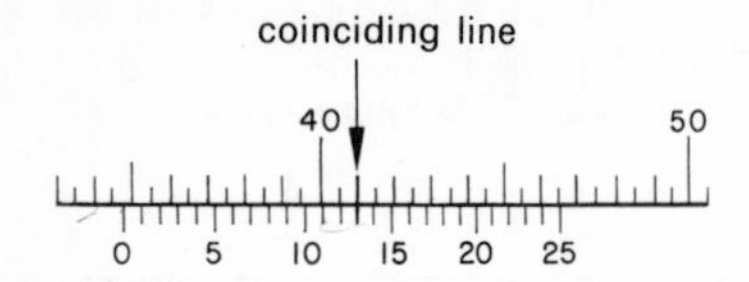

Figure 58 *An example of a reading on a 0.02 mm vernier scale*

ment is given by adding the main scale reading to the vernier scale reading.

Figure 58 gives an example of a 0.02 vernier reading. The reading on the main scale up to the zero on the vernier scale is 34.5 mm. On the vernier scale, the thirteenth graduation coincides with a graduation on the main scale. This represents a distance of
13×0.02 mm $= 0.26$ mm

$\therefore$ total reading $= 34.5 + 0.26$ mm
$= 34.76$ mm

Example 1

Figure 59 shows three 0.02 mm vernier scale readings; what measurements do they represent?

(a) The main scale reading is 77 mm.
The corresponding number of the vernier scale is 20.
$\therefore$ vernier reading $= 20 \times 0.02 = 0.4$ mm
$\therefore$ measurement $= 77.40$ mm

(b) The main scale reading is 14.5 mm.
The corresponding number of the vernier scale is 8.
$\therefore$ vernier reading $= 8 \times 0.02 = 0.16$
$\therefore$ measurement $= 14.66$ mm

(c) The main scale reading is 39.5 mm.
The corresponding number on the vernier scale is 12.
$\therefore$ vernier reading $= 12 \times 0.02 = 0.24$ mm
$\therefore$ measurement $= 39.74$ mm

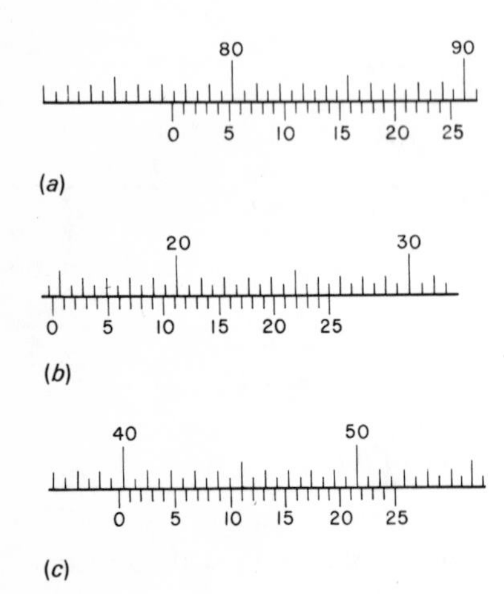

Figure 59 *Vernier scale readings*

When taking measurements using vernier calipers it is important to set the caliper faces parallel to the surfaces across which measurements are to be made. Failure to do so will result in an incorrect reading. Internal dimensions can be measured using vernier calipers; an example is shown in Figure 60. The reading given is for the gap between the inside faces of the calipers. For an internal measurement the sum of the thickness of the two jaws ($2L$ on Figure 60) must be added to the reading of the instrument. The distance L is usually clearly engraved on to the jaws of a vernier caliper.

Comparison of the micrometer and vernier calipers

In certain respects vernier calipers are more adaptable than micrometers; they can operate over a much larger measurement range that would require several micrometers; they can measure internal and external dimensions using the same measuring instrument. On the other hand, a micrometer is more robust, and is not prone to the same degree of distortion as vernier calipers. The barrel protects the

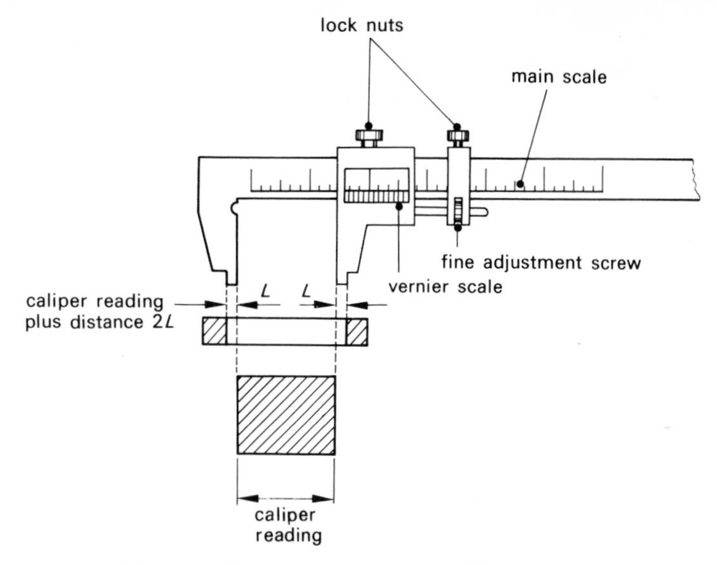

Figure 60 *Vernier calipers measuring internally and externally*

micrometer screw from damage, and also from dirt and small pieces of swarf, whereas the moveable scale of the vernier is unprotected.

A micrometer has divisions representing 0.01 mm units, but these can be subdivided by eye, so that the smallest measurement which can be conveniently estimated is approximately 0.005 mm. A vernier scale canot be read to measurements smaller than those signified by the divisions. Thus, a 0.02 vernier scale can be read to 0.02 mm.

The vernier depth gauge

Figure 61 shows a vernier depth gauge in use. The vernier scale is fixed to the main body of the depth gauge, and is read in the same way as vernier calipers. Running through the depth gauge body is the main scale, the end of which provides the datum surface from which measurements are taken. Care must be taken to ensure that this end datum surface just touches the surface at the base of the recess, while the measuring face remains in contact with the workpiece.

The vernier height gauge

The vernier height gauge (Figure 62) makes use of a longer vernier scale than has previously been discussed. Graduations on the main scale are in 1 mm intervals, but the vernier scale is equally divided into 50 divisions over 49 mm length: $\frac{49}{50}$ mm (or 0.98 mm) is the width of one division on the vernier scale. Note that the dimensional

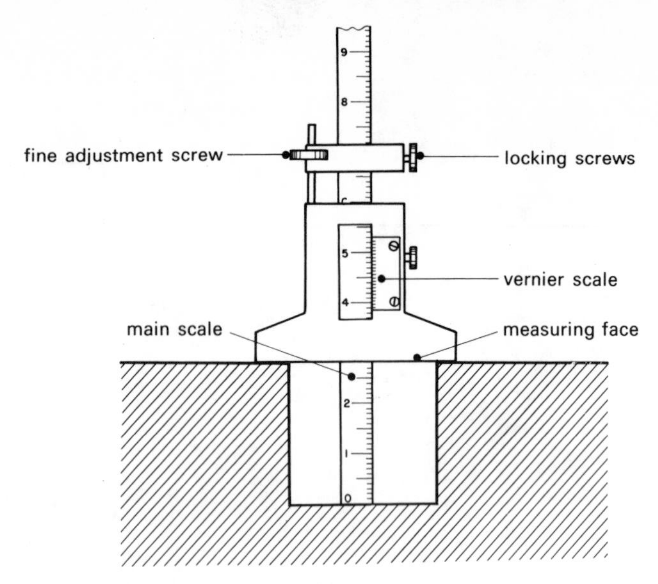

Figure 61 *Vernier depth gauge*

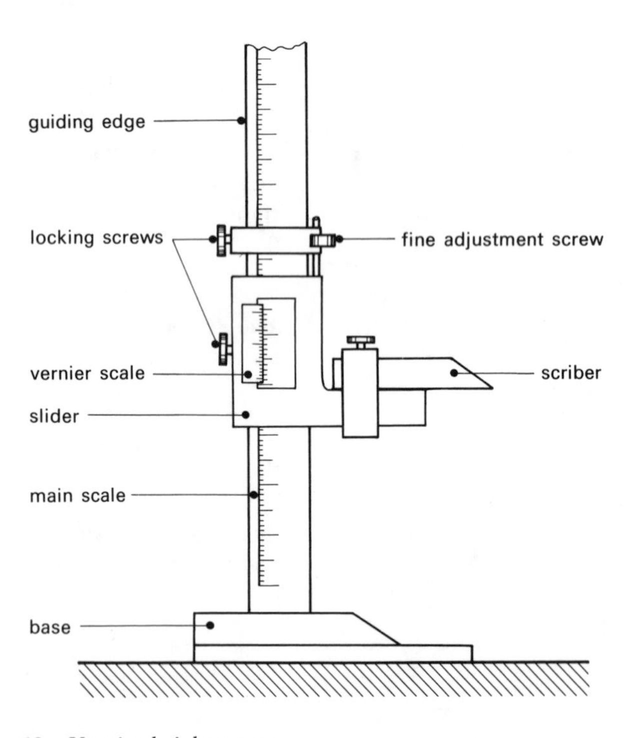

Figure 62 *Vernier height gauge*

accuracy of the instrument is the same as the previously described 12 mm vernier scale, but by increasing the scale length to 49 mm, greater ease in taking readings is obtained. This reduces the possibility of errors being made in the readings. The vernier scale is fixed to a slider which has a scriber attached to it (Figure 62). In this form the vernier height gauge is suitable for marking out operations requiring a high degree of accuracy.

The vernier protractor

The vernier protractor makes use of the same basic principle as the other vernier scales already described. On the main scale the graduations are in 1° intervals of an arc (Figure 63). The vernier scale occupies 46° on the main scale and is divided into 24 equal divisions, 12 on each side of a centre zero. As each side of the centre zero is marked from 0 to 60 minutes of an arc; then one of the divisions is equivalent to $\frac{60}{12}$ minutes, i.e. 5 minutes of an arc on the scale reading.

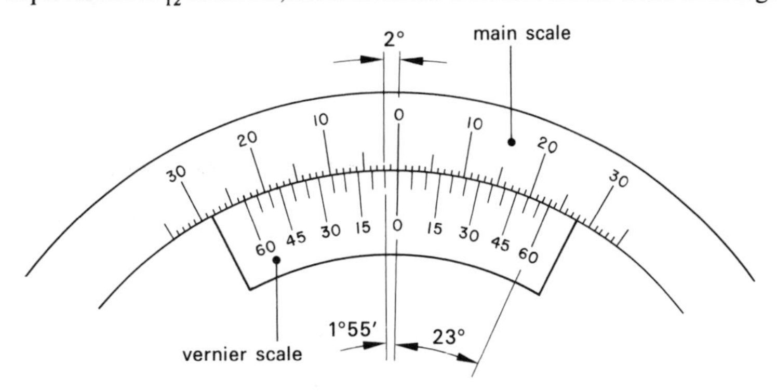

Figure 63 *The principle of the vernier protractor*

Since 12 divisions on the vernier scale occupy 23° on the main scale, each vernier division must be equal to $\frac{23}{12}^\circ = 1\frac{11}{12}^\circ$. Thus the smallest measurement which can be conveniently read is $2^\circ - 1\frac{11}{12}^\circ$, i.e. $\frac{1}{12}^\circ$ or 5 minutes.

To read the vernier protractor, the number of whole degrees on the main scale up to the vernier's centre zero mark is noted. Next, the mark on the vernier scale which is adjacent to a mark on the main scale is found, and this is added to the number of whole degrees to obtain the reading.

Figure 64 gives an example of a vernier protractor reading. The whole number of degrees before the centre zero mark on the vernier scale is 57°. The line coincident with a main scale graduation is shown as 10 minutes. Hence the reading is 57° 10 minutes.

A vernier protractor is a versatile instrument when used for measuring angles between faces on a component or workpiece. It is capable of measuring from 0° to 360°, when used in conjunction with an acute angle attachment. Care must be taken with the instrument, particularly the blade, which has a ground surface, if accurate readings are to be made.

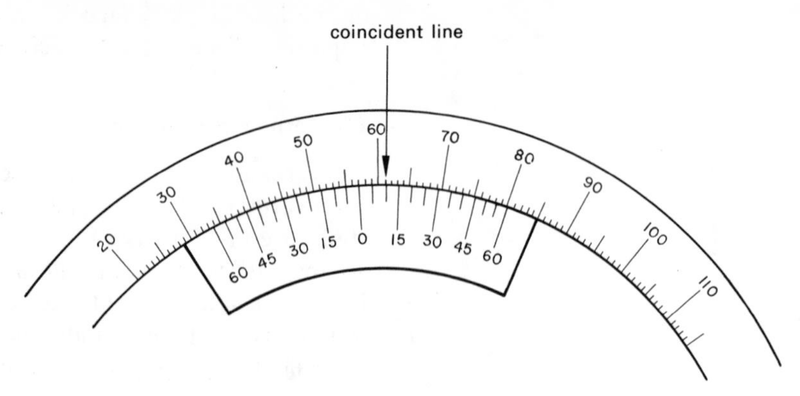

Figure 64 *Example of a vernier protractor reading*

Self-assessment questions

31 Draw a diagram which indicates a reading of 5.38 mm upon
(i) a micrometer
(ii) a vernier scale

32 Describe the principle of the vernier, using sketches where appropriate.

33 Vernier calipers are used to measure two dimensions:
(a) a 13 mm diameter bar and
(b) a 13 mm diameter hole.
If the thickness of each jaw is 3 mm, what does the vernier scale read for each dimension?

34 What is the main use of a vernier height gauge?

35 What is the smallest measurement which can be conveniently read with a vernier protractor?

36 Describe the principle of the vernier protractor.

37 List the advantages and disadvantages of:
(i) a micrometer and
(ii) vernier calipers,
under the headings of
(a) robustness,
(b) ease of reading of the instrument,
(c) adaptability,
(d) protection of the measuring device from dirt and accidental damage.

Reading the instruments

A micrometer has divisions representing 0.01 mm, but, with care, a division can be subdivided by eye, so that it is possible to measure a dimension within limits of ± 0.005 mm.

A vernier scale cannot be read to a measurement smaller than that signified by the divisions. Thus, using a 0.02 mm vernier scale, a dimension can only be measured to an accuracy of 0.02 mm, i.e. ± 0.01 mm.

The dial gauge indicator

Both micrometer screw and vernier scale instruments are capable of giving direct readings. There is, however, another range of instruments used in the measurement of components; they are collectively called *comparators*. Although not strictly a measuring instrument, a comparator is a useful instrument on many occasions. One such comparator is a dial gauge indicator (sometimes referred to as a 'clock gauge').

The principle of a comparator is to compare an unknown dimension with a known dimension. In doing so the dimension sought is not directly obtained; instead the limits of the required dimension are found.

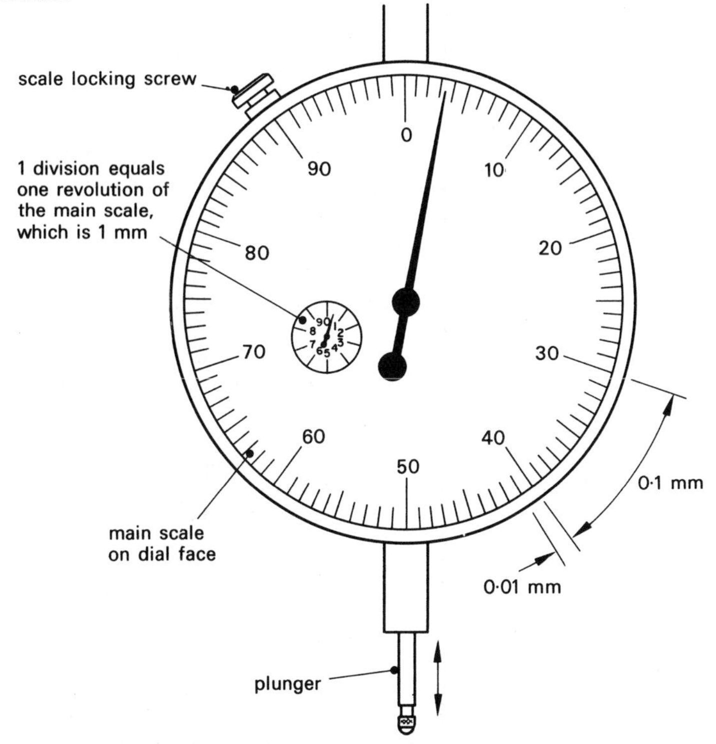

Figure 65 *Plunger-type dial gauge indicator*

Solutions to self-assessment questions

31 Readings on a micrometer and vernier are shown in Figure 66.

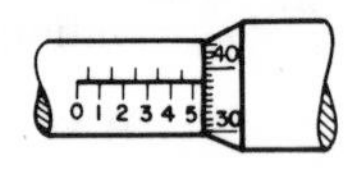

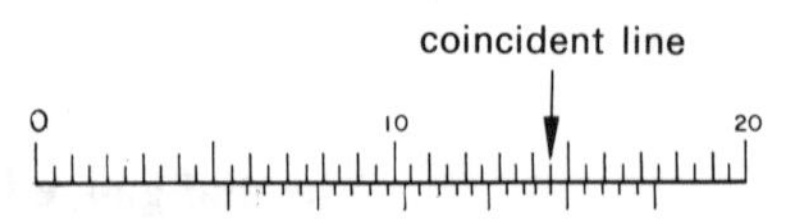

Figure 66 *Solution to self-assessment question 31*

32 The principle of the vernier is illustrated in Figure 56(a) (page 47), which contains a scale of length 9 mm divided into 10 parts. The vernier scale is read in conjunction with the main scale, which is marked in divisions of 1 mm; the vernier scale is marked in divisions of $\frac{9}{10}$ mm (i.e. 0.9 mm). Thus it is possible to read the scale to $(1.0 - 0.9)$ mm or 0.1 mm. The accuracy of reading of the scale can be increased by increasing the length of the vernier scale, a typical size being 12 mm divided into 25 graduations. The main scale graduations may also be changed from 1 mm to 0.5 mm. The smallest measurement which may then be conveniently read is:
$(0.5 - \frac{12}{25})$ mm $= (0.5 - 0.48)$ mm $= 0.02$ mm

33 (a) 13 mm
(b) 13 mm $- (2 \times 3)$ mm $= 7$ mm

34 The main use of a vernier height gauge is to measure or mark out components that require a high degree of dimensional accuracy.

35 A vernier protractor can be read to 5′ of an arc.

36 On a vernier protractor, the main scale graduations are in 1° intervals, with the vernier scale covering 46 of these graduations. If the vernier scale is divided into 24 equal divisions, then $\frac{46}{24} \div 1\frac{11}{12}^\circ$ or 1° 55′ is equal to one division on the vernier scale.

37 (a) Robustness. Owing to its shape and construction, the micrometer is stiffer and more robust than the vernier.
(b) Ease of reading. The majority of people find that a micrometer scale is slightly easier to read than a vernier scale. A small magnifying glass assists in determining the coincident lines on the vernier and main scales.
(c) Adaptability. Vernier calipers can be used over a much larger measurement range than a single micrometer; for example, if three dimensions are to be measured – 23.5 mm, 48.6 mm and 61.7 mm – three micrometers (0–25 mm, 25–50 mm and 50–75 mm) are required; only a single set of vernier calipers is required.
Internal and external diameters can be measured using a single set of vernier calipers, but inside and outside micrometers are required in order to make the same measurements.
(d) Protection from dirt and damage. The measuring screw of a micrometer is protected from dirt and accidental damage by the barrel, but the moveable and main scales of vernier calipers are unprotected.

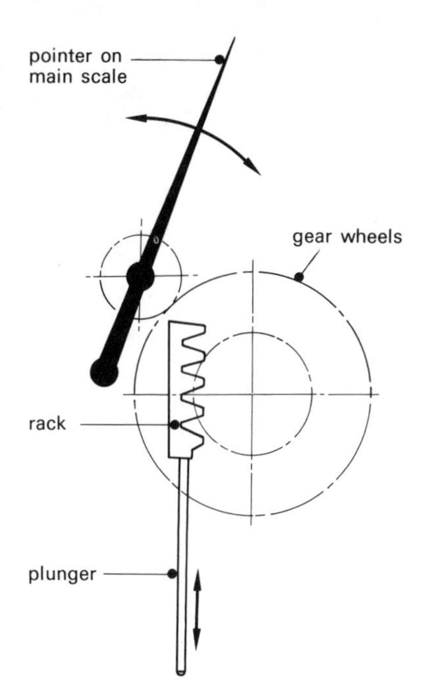

Figure 67 *Principle of operation of a plunger-type dial gauge indicator*

Figure 65 shows the main features of a plunger-type dial gauge indicator. The main scale is graduated into equal divisions corresponding to a 0.01 mm movement of the plunger. A second but smaller dial is set in the main dial face to indicate the number of complete revolutions turned through, one revolution beng equivalent to 1 mm of plunger movement. To enable the instrument to be set at zero for any convenient position, the main scale can be rotated and locked into place, using the scale locking screw indicated in Figure 65.

The principle of operation of the instrument is shown in Figure 67, when it can be seen that the plunger is attached to a rack. Meshing with a gear wheel, the straight or linear motion of the rack is converted into an angular or turning motion, the movement being magnified by using a large gear in mesh with a small gear wheel. It is the small gear wheel that is fitted to the main scale pointer shown in Figure 65.

The mechanism described above is simple, reliable and very sensitive. However, this sensitivity means that great care must be exercised when using a dial gauge indicator.

Figure 68 shows a dial gauge indicator fitted to a pillar stand; the stand enables the indicator to be locked in any required position.

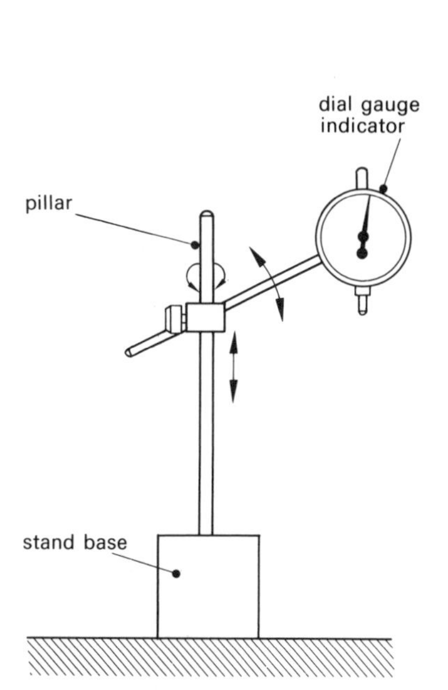

Figure 68 *Dial gauge indicator mounted on a pillar stand*

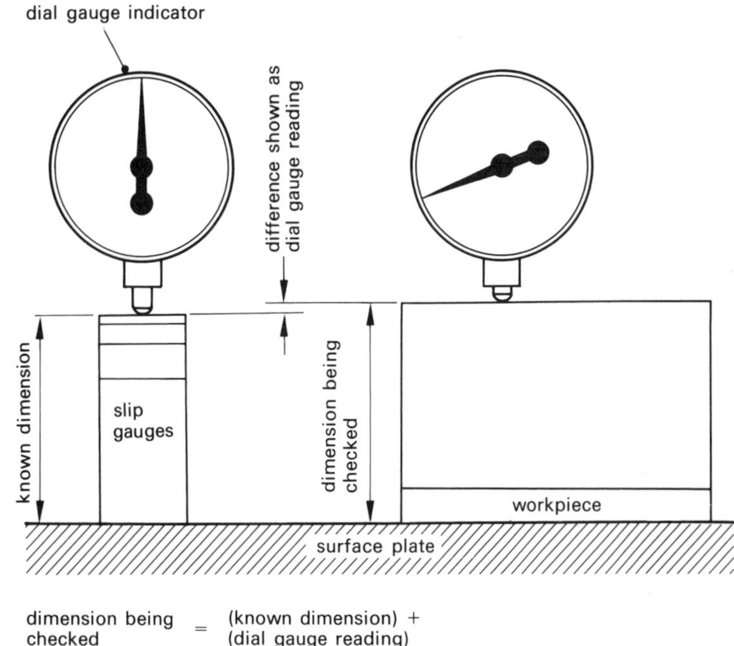

Figure 69 *Plunger-type dial gauge indicator in use*

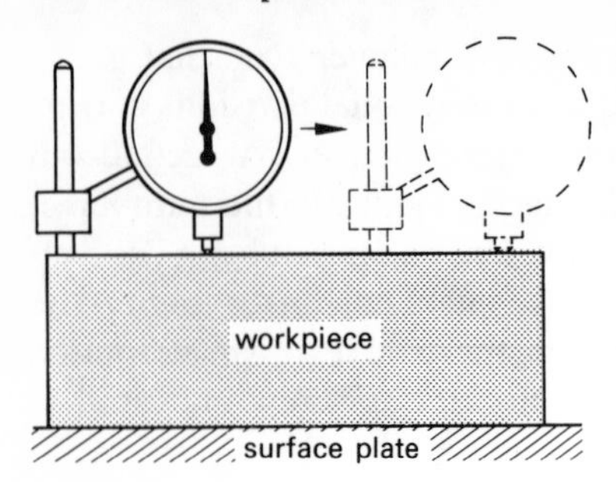

Figure 70 *Using a dial gauge indicator to check for parallelism*

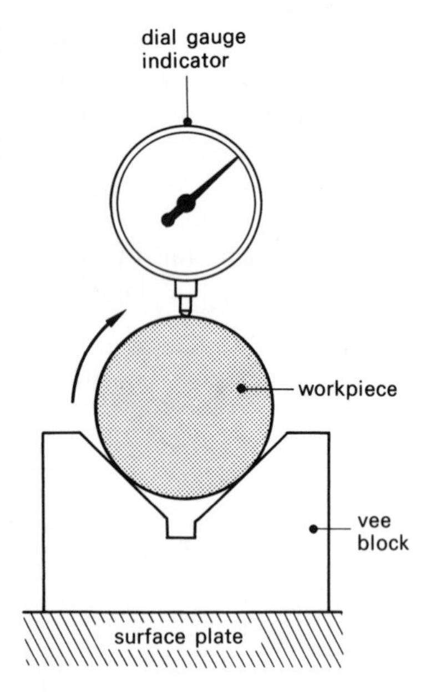

Figure 71 *Using a dial gauge indicator to check the roundness of a component*

Figure 69 shows a workpiece being compared with a known standard, which may be, for example, slip gauges. With the dial gauge indicator placed so that the plunger is in contact with the reference face, the dial scale can be set to zero and locked. Moving the pillar stand (with the dial gauge indicator fitted to it) across to the workpiece, which is positioned on the same flat surface, any difference between the original setting and the face being checked can be read off the main scale.

Faces on a workpiece that are required to be parallel to one another can be checked as shown in Figure 70. The dial gauge is set at one end of the component and the reading noted. The plunger is then moved across the face of the workpiece; if the two faces are parallel, the reading does not alter.

It is not only flat surfaces that can be checked using a dial gauge indicator. Figure 71 shows how a component may be checked for roundness. Placing the workpiece in a vee block and bringing the dial gauge plunger into contact with the upper face, a datum reading can be taken. If the workpiece is now slowly rotated in the vee block, the dial gauge pointer remains stationary for a truly round surface. Movement of the pointer indicates that the periphery of the component is not truly round.

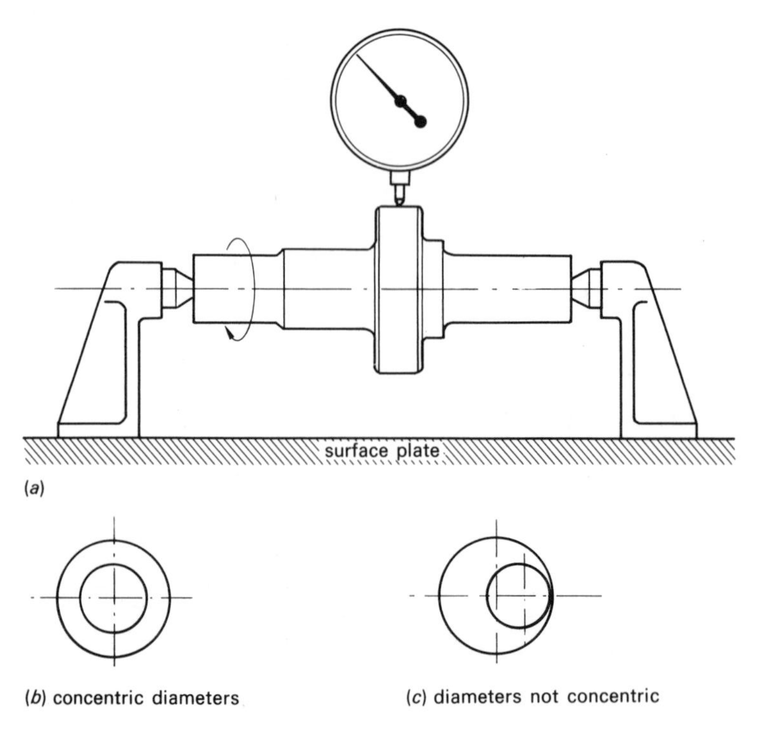

Figure 72 *Using a dial gauge indicator to check for concentricity*

When it is required that the diameters on a shaft are concentric (that is, they have the same central axis), one method of checking is shown in Figure 72(a). The workpiece is mounted between centres, which lie on the centre line of the component. Bringing the plunger of the dial gauge indicator into contact with each diameter in turn, the workpiece can be slowly rotated. For true concentricity, the pointer on the main scale remains stationary, showing that each part of the surface of the component is an equal distance from the main central axis.

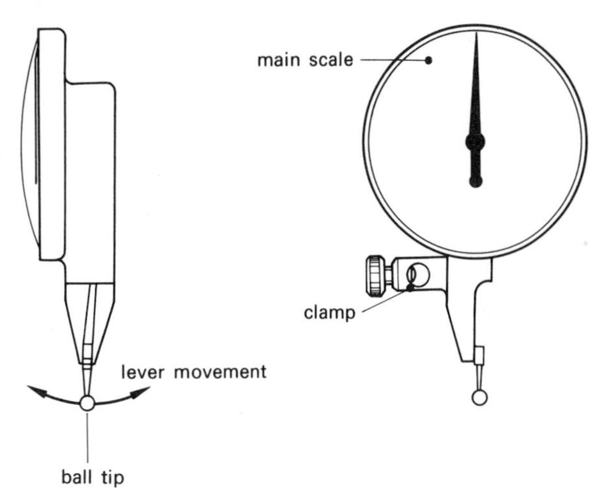

Figure 73 *Lever-type dial gauge indicator*

Figure 73 shows a lever-type dial gauge indicator in which the plunger has been replaced by a ball-tipped lever arm that is pivoted on the body of the indicator. The movement of the lever is such that it is at right angles to the main scale. A combination of the compactness of the design and the relative direction of lever movement gives the instrument its advantages. With the ball tip placed inside a small bore as illustrated in Figure 74, concentricity, roundness or machine settings can be checked – operations that are difficult to carry out using a plunger type of dial gauge.

The disadvantage of the lever dial gauge is that it has a small measuring range. Figure 75 shows the principle of operation of the instrument. The ball-tipped lever is connected to a main lever, which is pivoted at its lower end. A ball at the upper end of the main lever engages in a scroll, so that movement of the lever rotates the scroll. It is this movement that is displayed by the pointer on the main scale, so enabling readings to be made. Using such a mechanism means that the ball tip movement is limited by the scroll length, which usually only allows about 1½ revolutions of the pointer to be made.

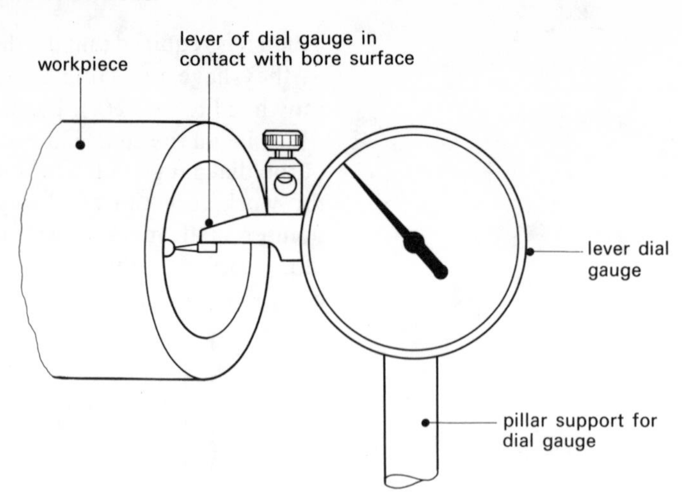

Figure 74 *Lever-type dial gauge indicator checking a small bore*

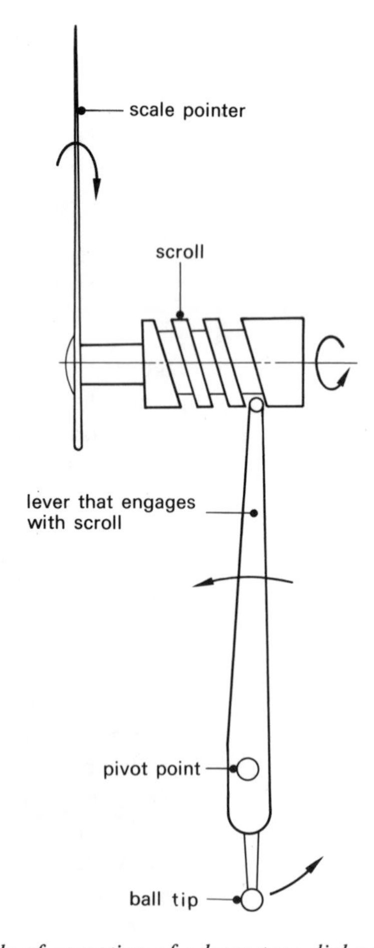

Figure 75 *Principle of operation of a lever-type dial gauge indicator*

Self-assessment questions

38 Explain what is meant by a 'comparator'.

39 By drawing a sketch, show a method of using:
(a) a dial gauge indicator to check the roundness of a solid bar 30 mm diameter
(b) a lever test indicator to check the concentricity of the bore of a hollow bar, 50 mm outside diameter and 25 mm inside diameter

40 Sketch a dial gauge indicator being used to check the parallelism of the machined surfaces of a steel component.

Solutions to self-assessment questions

38 A comparator is an instrument that compares or indicates the difference between measurements. Because a comparator has a small measuring range, the comparison must be made so that any differences between the datum and the workpiece are small. See Figure 69 for an example.

39 (a) A method of using a dial gauge indicator is shown in Figure 71 (page 56).

(b) The concentricity of the bore can be checked by a method similar to that shown in Figure 72 (page 56).

40 A method of checking for parallelism is shown in Figure 70 (page 56).

Topic area: Machine tools

After reading the following material, the reader shall:

5 Know the method of producing holes using sensitive drilling machines.

5.1 Identify the features of a sensitive drilling machine such as table, column, spindle head.

5.2 Identify the features of twist drills, trepanning tools and reamers.

5.3 Recognize the need for clamping from the point of view of operator safety.

5.4 Explain the problems associated with producing holes in sheet metal, such as grabbing, lobed holes, etc.

In engineering, 'drilling' means making a hole where none previously existed or enlarging a small hole to a larger diameter. The hole is bored out by a hardened tool which is driven by a machine called a drilling machine. Drilling is the most common method of making holes. It is quite satisfactory in many instances, and can be carried out at low cost. However, compared with many other metal machining processes, the finish produced by drilling operations is fairly rough, and both the accuracy of the hole size and the location may be poor.

The drilling of holes is an essential feature in engineering production. There are very few engineering components which do not have holes drilled in them somewhere. Yet drilling – a seemingly simple operation – is given little prominence in the study of engineering development. Drilling appears deceptively simple, yet the production of satisfactory holes is one of the most difficult metal removing processes.

Ever since man began to lead a civilized life, he has needed to make holes in materials. To produce holes in animal skins he used sharpened pieces of bone. The oldest form of drill was probably invented in late Palaeolithic times – perhaps 10 000 years ago. This was the bow drill. The string of a small bow was given a turn about a rod-shaped flat bit which was then rotated as the bow was moved briskly to and fro like a saw. The bit was sometimes sharpened, but was more often assisted in its cutting action by the use of sharp sand or other abrasives. About 3000 years ago, the Egyptians were using bow drills, but technological developments by then enabled them to

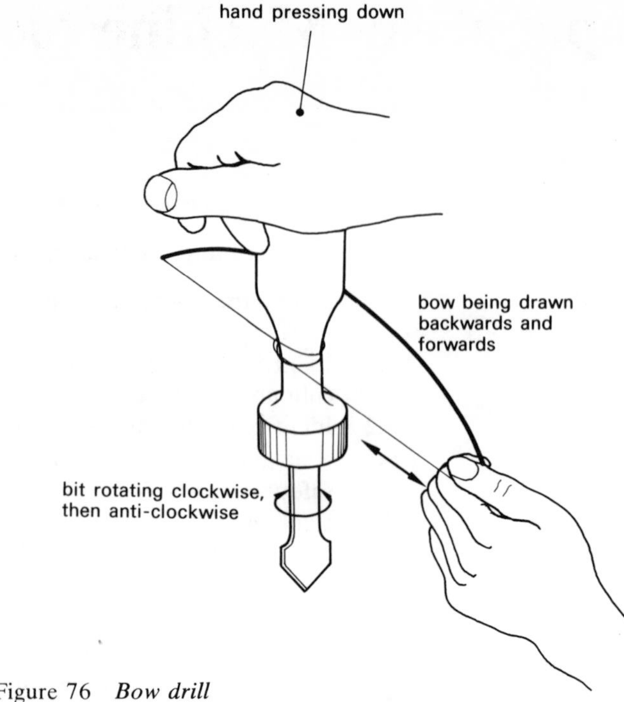

Figure 76 *Bow drill*

use a bit made of copper; this copper bit was hardened by hammering. The bow drill has continued in use, almost unaltered, up to the present day. It has been employed recently by jewellers, and by people repairing chinaware.

Progress in the design of all machine tools was slow for thousands of years. Simple drilling machines were in existence during the seventeenth century, but it was not until the first industrial revolution – about 200 years ago – that drilling operations took on the pattern that they have today. It is really only in the last 150 years that drilling machines became relatively sophisticated pieces of equipment.

The modern drilling machine consists of a base which supports a vertical column which carries

(i) a table to support the workpiece being drilled
(ii) an arm provided with bearings for the drilling spindle

In most machine tools the cutting tools can be supported close to their operating edge. However, in drilling, there is always a large overhang at the cutting area; in consequence the twist drills used have to be both elastic and tough. This reduces the cutting efficiency of a drill.

Most machine tools appear to be a complex mixture of intricate parts. However, seemingly complicated patterns can frequently be analysed into a number of fairly simple components. Most machine tools – including drilling machines:

(i) Hold the workpiece which is to be shaped or cut.
(ii) Hold the cutting tool.
(iii) Move (a) the workpiece, or (b) the cutting tool, or (c) both the workpiece and the cutting tool.
(iv) Provide a feeding movement for the workpiece or the cutting tool or both.

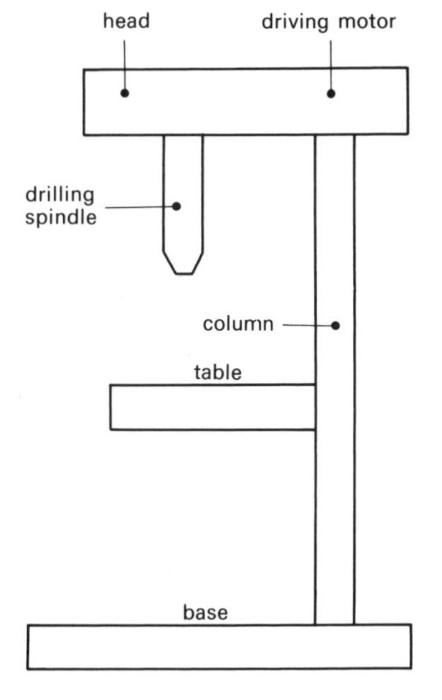

Figure 77 *Basic elements of a drilling machine*

All drilling machines have some means of rotating the cutting tool, this tool being advanced along its own axis into a stationary workpiece.

The *sensitive drilling machine* (Figure 78) is a relatively light form of machine used to drill holes no larger than about 12 mm diameter. In this type of machine the drill is fed into the work by a hand lever attached directly to the drilling spindle. It is because the machine operator is directly sensitive of the cutting action of the drill that these machines are called sensitive drilling machines.

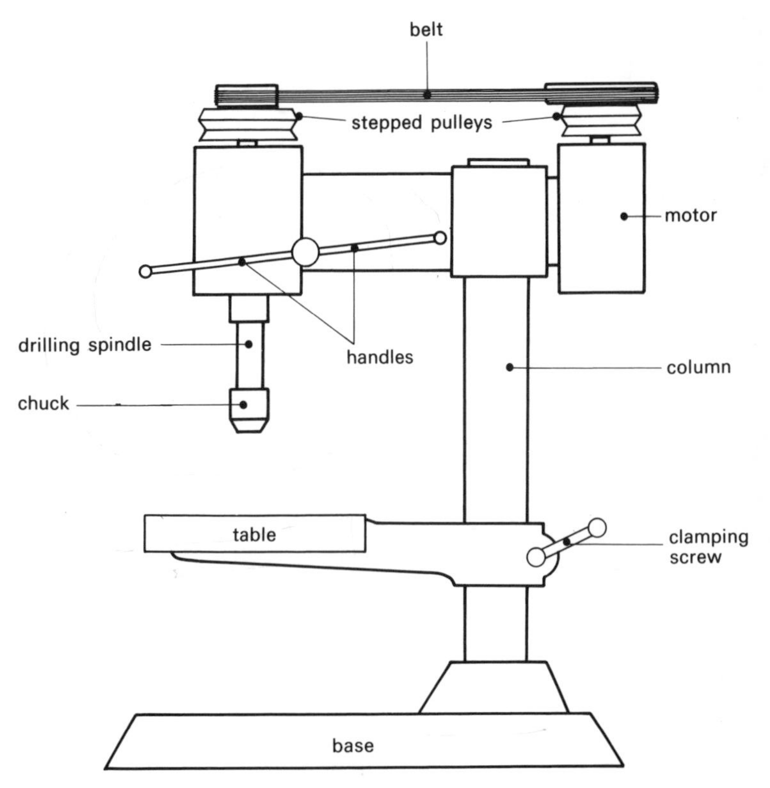

Figure 78 *Sensitive drilling machine*

The speed of rotation of the spindle driving the drill can be altered by moving the belt to a different pair of the stepped pulleys located at the top of the drilling machine. Compared with larger drilling machines, the sensitive drilling machine has a rather restricted range of speeds. The speeds at which drills are driven in sensitive drilling machines are much higher than the speeds in larger drilling machines.

The work table of a sensitive drilling machine is a plane surface. It acts as a locating face for the work to be drilled. To avoid errors in drilling, it must be used carefully and kept clean. The work table can be raised or lowered by slackening the clamp screw securing it to the column, while on some machines it is possible to revolve the table on the support collar (Figure 79).

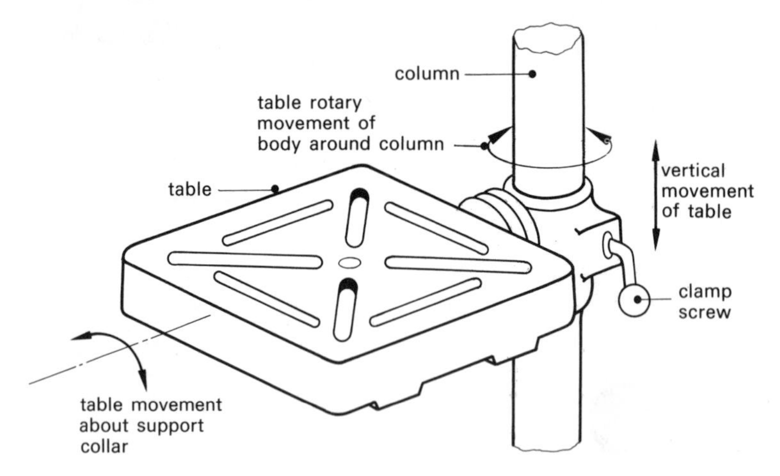

Figure 79 *Movement of the table of a sensitive drilling machine*

The twist drills commonly used in workshops are a direct development of the flat drills used for many years before the development and introduction of power-driven tools and new materials. The modern drill is a tool with cutting edges at one end, and it has flutes which allow the release of the chips cut from the workpiece.

Drills are produced either with (i) straight shanks, or with (ii) taper shanks.

(i) Straight (or parallel) shank drills (Figure 80(a)) depend on friction between the shank and the jaws of the chuck to rotate the drill as it cuts into the workpiece. Straight shank drills are frequently used in sensitive drilling machines. These drills depend for their accuracy on the chucks which hold them. High-quality chucks are initially very accurate, and the small drills commonly used in sensitive drilling machines can be relied upon to give good results. Unless the chuck is in good conditon, the drilling may lose accuracy, (a) dimensionally (the hole may not be round) and (b) positionally (the hole may not be in the required position).

(ii) Whilst straight shank drills are fitted into a chuck on the drilling machine, taper shank drills (Figure 80(b)) are fitted directly into the drilling machine spindle. Taper shank drills have a positive, mechanical drive because the tang of the drill is located in a slot. Taper shank drills normally produce high-quality results, provided that the taper shanks are maintained in good condition.

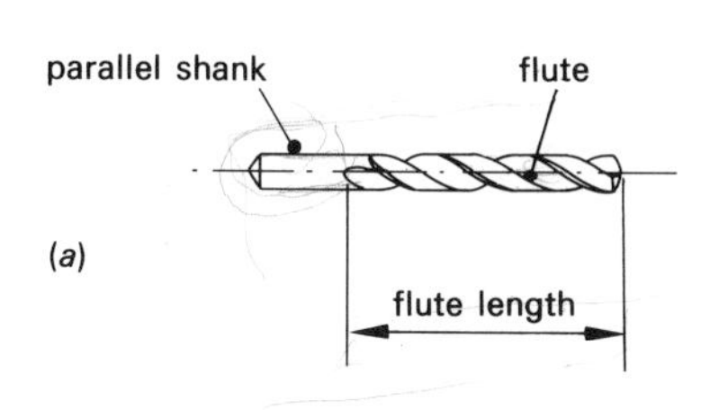

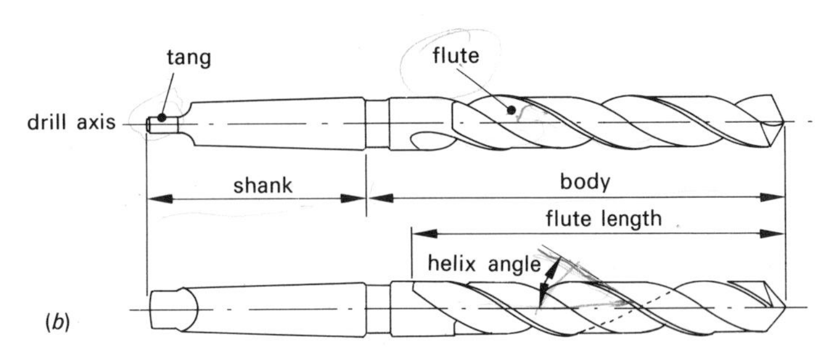

Figure 80
(a) *Parallel shank drill*
(b) *Taper shank drill*

The success of any drilling operation depends to a great degree on the accuracy with which the drill has been produced. Although the drill appears to be a fairly simple tool, its appearance is deceptive. For the efficient cutting of any particular material, the helix angle and the point angle are specified.

As the drill moves through the workpiece it is important that the process be efficient. Rubbing between the drill and the workpiece wastes energy and generates undesirable heat. In order to reduce friction, the drills are ground so that the backs of the cutting edges slope away from the workpiece.

The drilling of long, relatively small diameter holes presents difficulties which prevent the use of ordinary drilling methods. Even if passages to take coolant to the cutting area are provided, overheating frequently occurs.

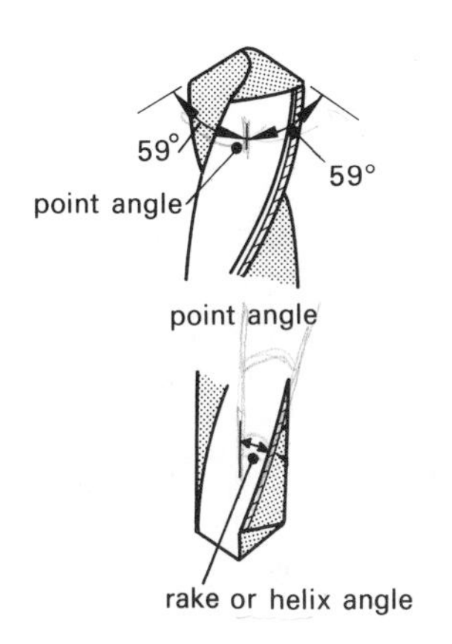

Figure 81 *Important angles on a twist drill*

Even on drilling machines much larger than the sensitive drilling machine the largest hole that can be drilled in normal circumstances is about 100 mm in diameter. When holes bigger than this are required, there are several methods that can be considered. One of these methods, *trepanning*, is frequently carried out on sensitive drilling machines.

The action of trepanning removes a core of material in one piece from the workpiece. Since the cores produced in trepanning operations are sometimes of a reasonable size, these cores can occasionally be used to manufacture small components.

A trepanning cutter has teeth at one end of a hollow cylinder. When machining, the cutter teeth make a groove in the workpiece allowing the solid core produced to pass inside the cutter. Trepanning is frequently employed to make holes between 60 mm and 220 mm diameter.

In passing, it is of interest to note that ancient Egyptian doctors as early as 1000 BC carried out operations on the skulls of patients, using trepanning tools to cut a disc of bone out of the top of the skull. From examination of mummies found in a good state of preservation it is clear that some patients lived for many years following a successful operation that must have been an appalling experience.

The prime function of a drill is to remove the maximum volume of material in as short a time as possible. However, the surface finish of the drilled holes is sometimes not of a high standard. To improve the surface of the drilled holes, a *reamer* can be used in a finishing process. There are two groups of reamers:

(i) those which are hand operated
(ii) those which are machine driven (Figure 82).

Considering group (ii), the machine reamer has a bevel lead, and the shank is usually tapered. The helix on a machine reamer is in the opposite direction to the helix on a twist drill. This is so even though they both are driven in the same direction during cutting. The 'left-hand' helix on the reamer prevents it being 'drawn-in' to the hole whilst it is being driven.

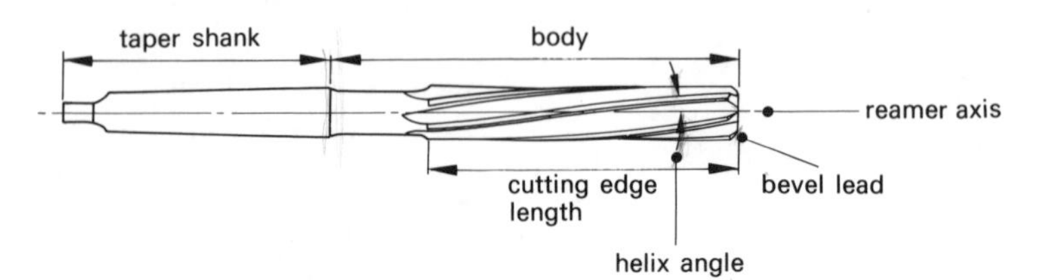

Figure 82 *Machine reamer*

Reaming is an operation by which holes are produced with both a good finish and an accurate size. Reaming is a scraping action more than a cutting action. Compared with other metal-machining tools, reamers have a comparatively large area of the tool in contact with the metal being removed. The best results are obtained with the reamer being driven at a fairly slow speed. Satisfactory use of reamers in sensitive drilling machines (which are driven at relatively high speeds) necessitates the supply of a generous flow of coolant.

Care must be taken when using all tools. A mishap with a hand tool such as a chisel may cause a bruise or even a cut, but an accident when using a machine tool may have much more serious consequences.

On drilling machines, accidents arise from three main sources:

(i) Tangs are twisted off.
(ii) Loose clothing or long hair becomes entangled in the drill.
(iii) The workpiece is not firmly located.

(i) The twisting off of tangs is still a serious problem for drilling machine operators. Frequently the tang fails because it has received rough and careless treatment during the removal of the drill from the spindle.
(ii) To prevent the operator being drawn into the drill, the appropriate guards around the spindle must be used at all times.
(iii) When carrying out a drilling operation it is important to ensure that the workpiece is firmly clamped in position. Figures 83, 84 and 85 show three examples of workpieces being held securely.

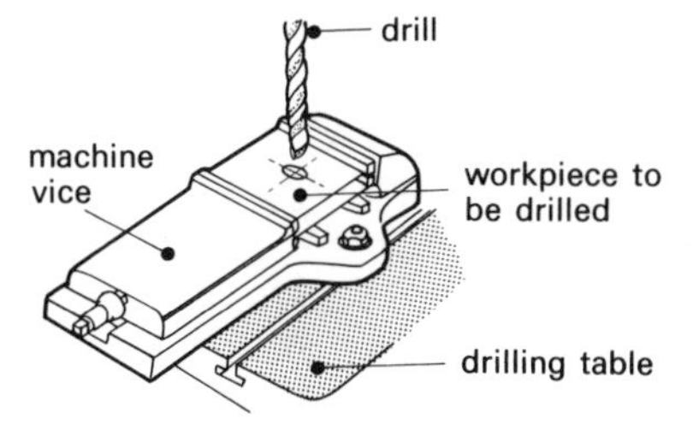

Figure 83 *Drilling machine and machine vice*

It cannot be stressed too strongly that material which is being drilled should *never* be held in the hands of the machine operator.

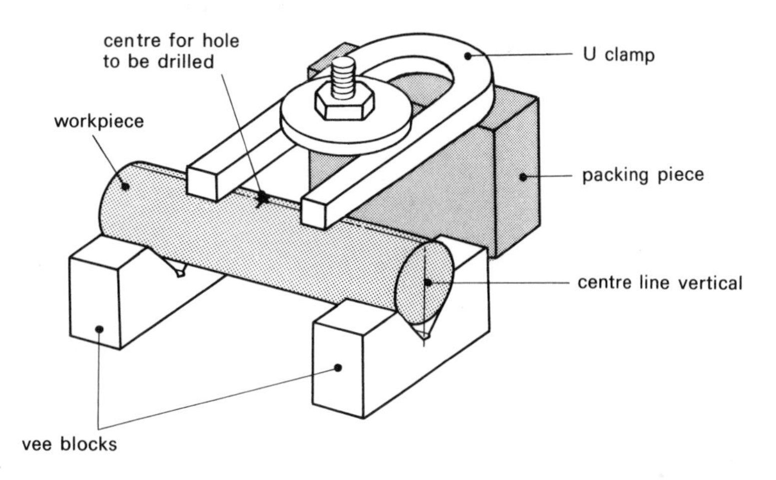

Figure 84 *Clamping a round bar for drilling*

Drilling thin metal sheets can be a dangerous procedure unless care is taken. The workpiece is readily lifted from the table by the drilling action and serious injuries can ensue. Since thin sheets cannot be held properly in a vice, they are frequently clamped to the table of the drilling machine. Care must be taken not to damage the table by drilling through the thin plate. In order to keep the metal sheet flat, and also to protect the work table, the sheet material is very often firmly clamped between flat pieces of wood.

When twist drills are used to produce holes in sheet material, they have a tendency to 'grab' at the metal and to dig into the periphery

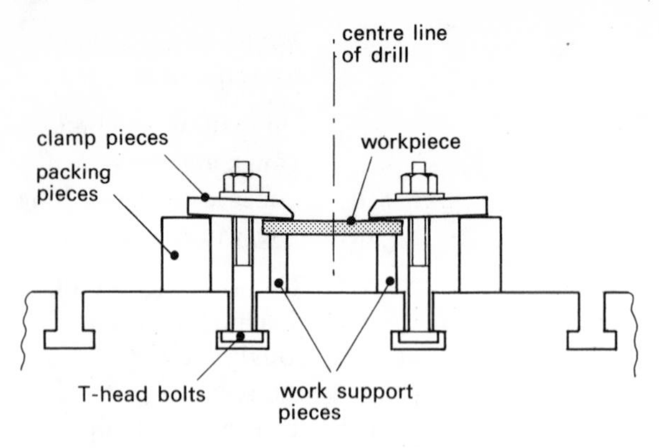

Figure 85 *Clamping flat bar for drilling*

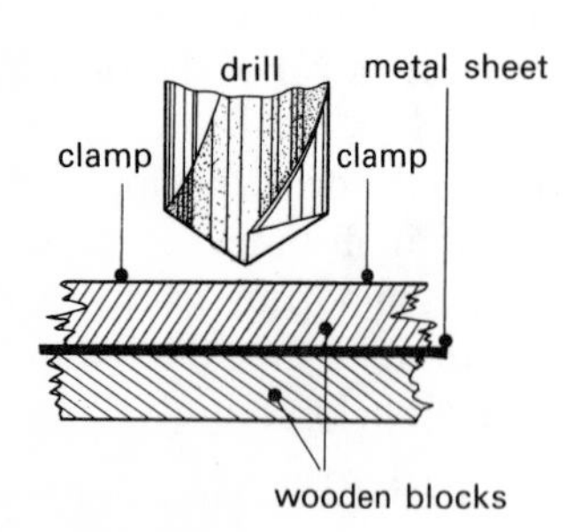

Figure 86 *Drilling large holes in thin metal sheet*

of the hole. Even when the twist drill makes a hole in a sheet metal, upon examination the hole is frequently found to be 'lobed' (shown somewhat exaggeratedly in Figure 87).

Holes can be produced more satisfactorily in sheet metal by using (i) saw drills or (ii) fly cutters. Both processes require that a small hole be made in the centre of the required hole, this providing guidance for the main cutter.

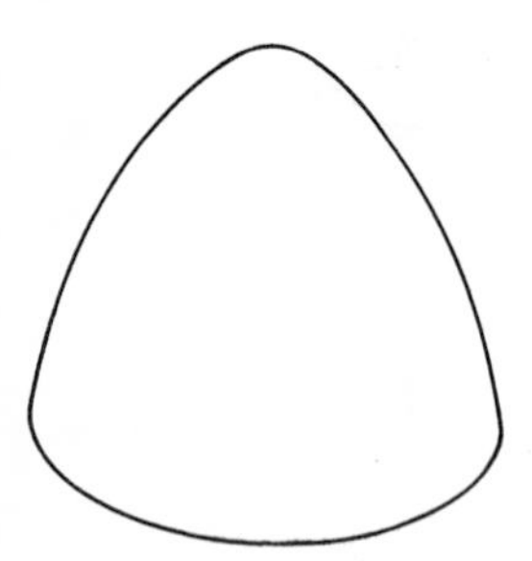

Figure 87 *Lobed drilled hole*

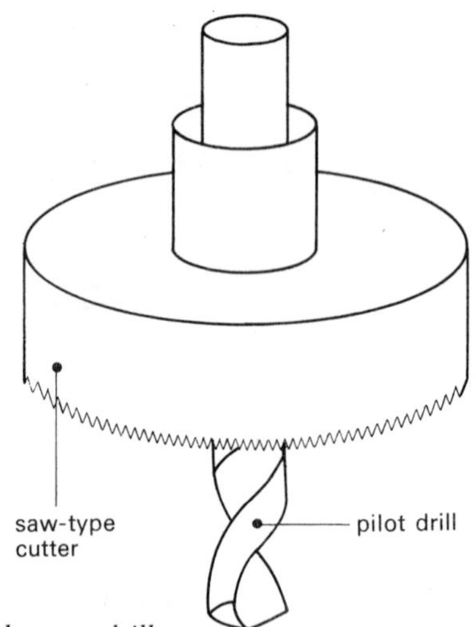

Figure 88 *Circular saw drill*

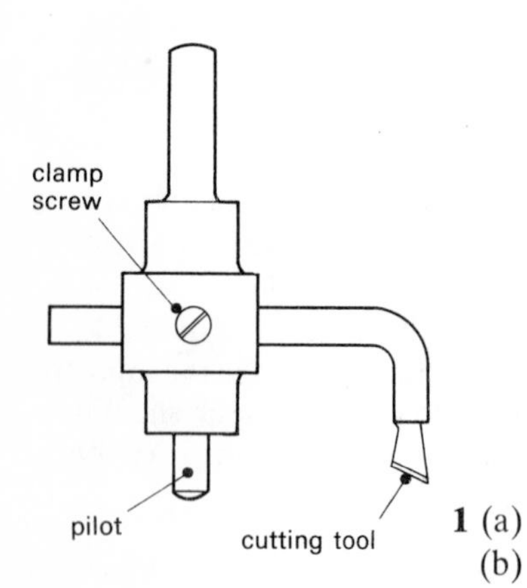

Figure 89 *Fly cutter*

Self-assessment questions

1 (a) Explain why the 'sensitive drilling machine' is so called.

(b) State two ways in which accidents can be avoided when using this type of machine tool.

2 With the help of Figure 90 complete the following table:

number	*name of part*
1	
2	
3	
4	
5	

3 With the help of Figure 91 complete the following table:

number	*name*
1	
2	
3	
4	

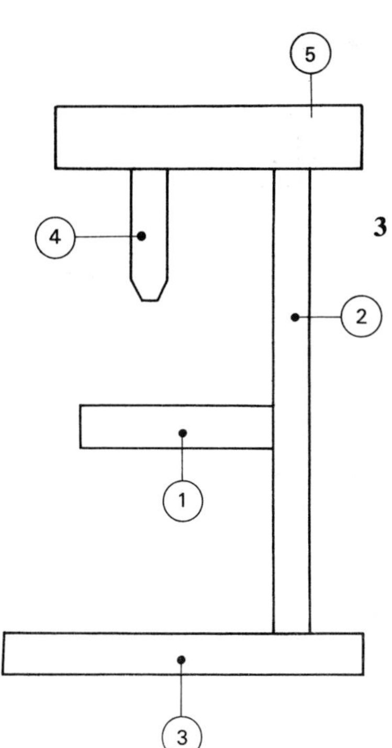

Figure 90 *Sensitive drilling machine*

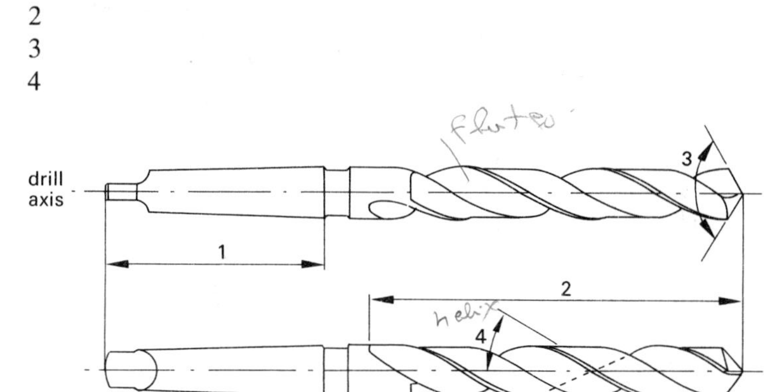

Figure 91 *Twist drill*

4 Name a method of producing holes which have diameters greater than 100 mm.

5 Reaming is used only as a finishing process.
TRUE/FALSE

6 A machine reamer must be driven at a fairly slow speed.
TRUE/FALSE

7 Why is it very desirable that a 'hand-feed' is used when drilling small diameter holes?

8 Why is it essential that the work table of a sensitive drilling machine be kept clean and free from damage?

9 Explain why work must be firmly clamped before drilling operations are started.

10 State two problems that may be encountered when drilling holes in thin metal sections.

11 Name two devices that may be used to produce holes in sheet material.

After reading the following material, the reader shall:

6 Know basic sheet metal operations.

6.1 Describe the use of manually operated guillotine, bending machine and hand tools to produce basic rectangular shapes.

Solutions to self-assessment questions

1 (a) In a sensitive drilling machine the drill is fed into the workpiece by a lever controlled by the hand of the person operating the machine. This method of operation brings into play the sense of 'feel' as the drill cuts into the metal. The name 'sensitive drilling machine' is derived from this fact.

(b) In any drilling operation accidents can be avoided by:

(i) Using only drills which have been carefully treated during operations to remove them from the drilling spindle. This will ensure that the tangs do not twist off.

(ii) Not wearing loose clothing or unrestrained long hair, and ensuring the use of appropriate guards.

(iii) Ensuring that the workpiece is secure before commencing drilling operations.

2

number	*name of part*
1	table
2	column
3	base
4	drilling spindle
5	driving motor

3

number	*name*
1	shank
2	flute length
3	point angle
4	helix angle

4 Trepanning.

5 True.

6 True. However, if a copious flow of coolant is used, the operation can be carried out on a sensitive drilling machine.

7 Because the cutting faces of a drill are always a distance from the point of support (the chuck or the spindle), drills can be deflected from the axis which they are required to follow. In spite of being made both tough and elastic, the smaller drills are very liable to break. A well trained and experienced operator can 'feel' when a drill is liable to fail, and can take appropriate action.

8 The work table acts as a locating face for the workpiece which is to be drilled. Any material coming between the workpiece and the table will cause problems in accurate drilling. The table must not be damaged (e.g. by drilling into it through the work-piece) or it will no longer be a useful datum surface.

9 During a drilling operation the drill may 'snatch' at the workpiece. This may cause an accident, damaging to the operator, the machine, or the workpiece. Firm location of the workpiece is therefore essential.

10 (i) The holes produced may be 'lobed' (i.e. not round).

(ii) The drill frequently 'grabs' at the workpiece causing inaccurate hole production.

11 (i) Fly cutter

(ii) Saw drill

6.2 Determine the required bending and folding allowances.

Sheet metal is used in the construction of a large variety of goods. For example, the bodies of washing machines, refrigerators and filing cabinets are made from sheet metal. Similarly, the outer shell or 'skin' of cars and aeroplanes is made from sheet metal; it is also used in the manufacture of ducting, such as is used to extract fumes from, or bring fresh air into a factory or office.

The body shell of a car is produced by manipulating sheet steel into relatively complex shapes by very large presses; the cost of the presses can be justified because of the great number of bodies which are produced. In contrast, ducting for ventilation may often be made by hand because ducting installations vary from factory to factory, or from one part of a factory to another.

Cutting operations

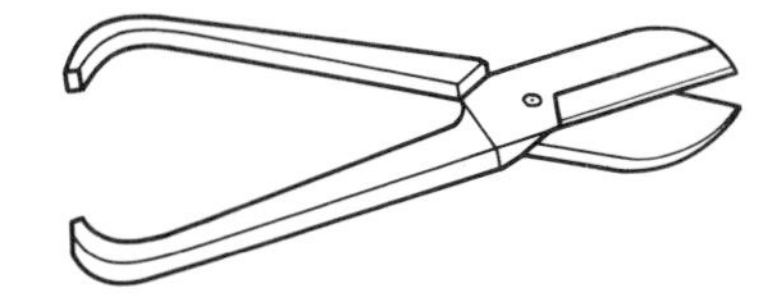

Figure 92 *Straight-type shears*

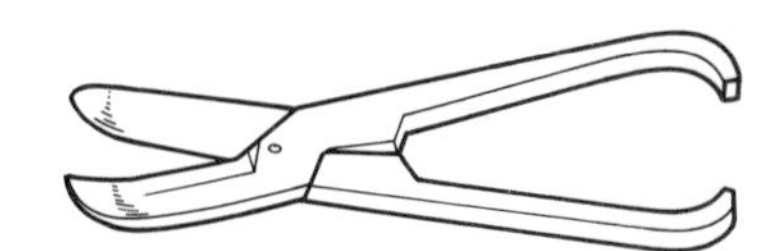

Figure 93 *Curved-type shears*

The term *sheet metal* is normally used to describe metal sheets with a maximum thickness of 2 mm. Above this thickness, it is usual to use the term *plate* metal.

Sheet metal can be cut by hand using shears or 'snips'. Snips are made either with straight blades (Figure 92) or with curved blades (Figure 93). Straight snips are used for cutting along straight lines and for trimming edges whereas rounded profiles can be produced using curved snips.

If a large straight length of sheet metal is to be cut, the treadle guillotine shown in Figure 94 is very useful.

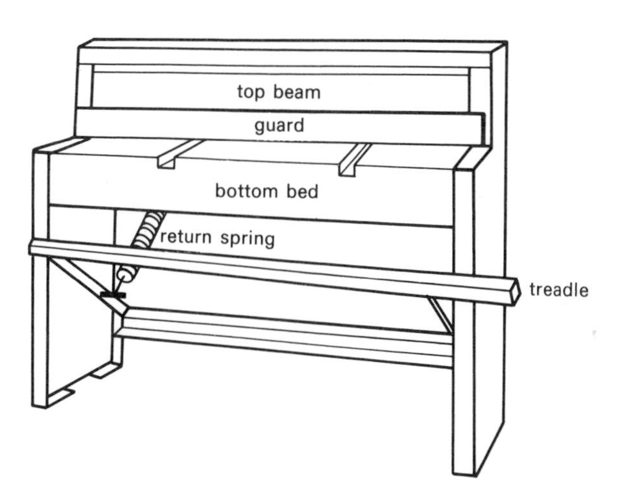

Figure 94 *Treadle guillotine*

The bottom bed and blade of the guillotine are fixed; the foot operated treadle activates the top blade, which moves vertically down to shear the metal in a continuous and progressive movement (Figure 95).

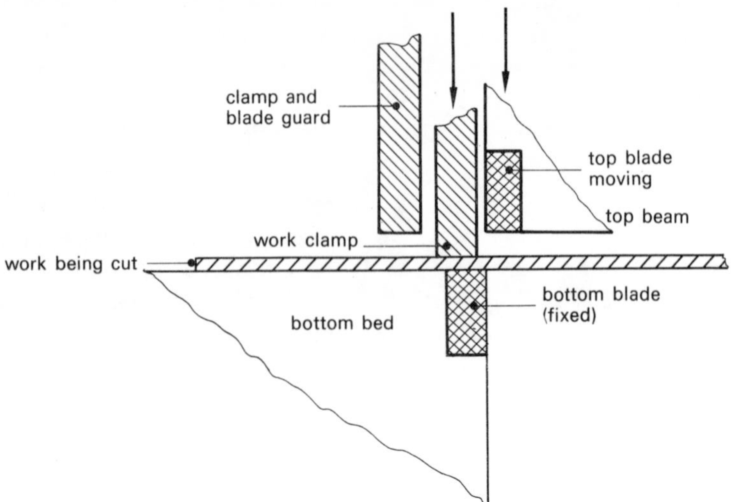

Figure 95 *How a guillotine works*

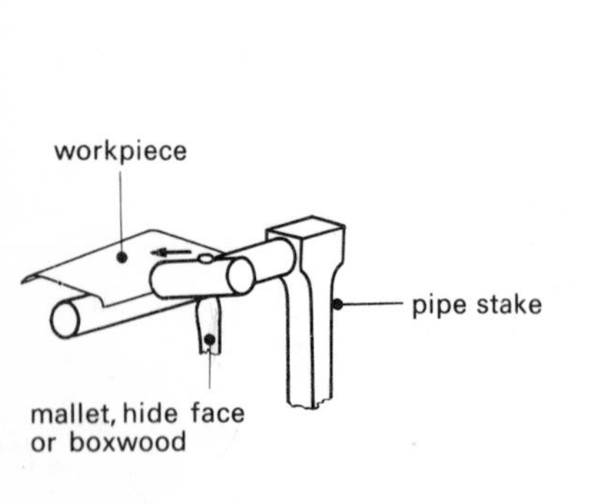

Figure 96 *Bending sheet metal with a hand tool*

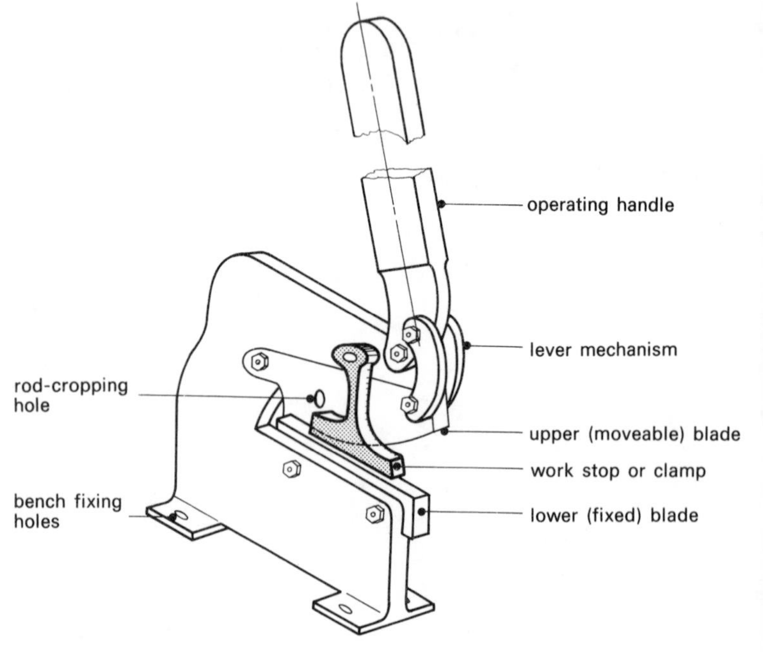

Figure 97 *Bench shears*

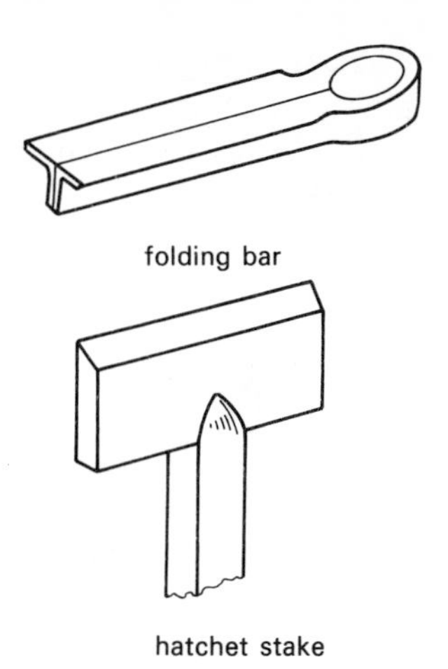

Figure 98 *Sheet-metal hand tools*

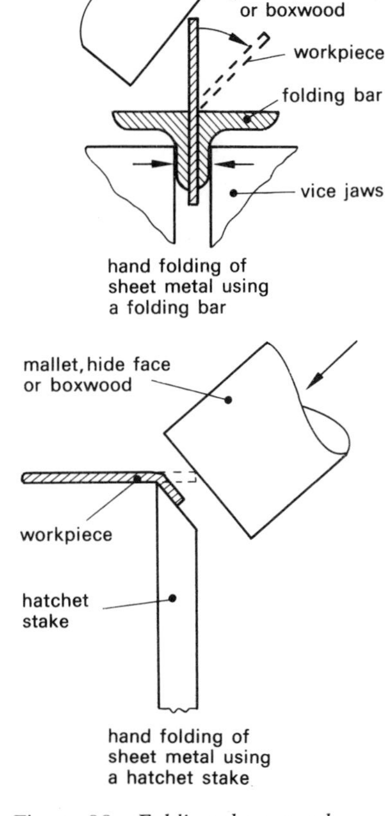

Figure 99 *Folding sheet metal by hand*

The guillotine shown in Figure 94 is limited to metal no thicker than 1.5 mm. For thicker sheets, bench shears can be utilized (Figure 97). The extra force required to cut the thicker sheet is obtained by a lever system connecting the operating handle and the upper (movable) blade.

Bending and folding operations

A number of methods can be used to bend or fold sheet metal by hand. In all the methods, it is important to bend the full width of the material steadily, so avoiding excessive local stretching or buckling.

The most common hand tool is the tinman's mallet; it has a box-wood or hide face in order to minimize damage to the metal. Dependent upon the type of bend or fold to be formed, a variety of tools or *stakes* can be used. Figure 96 shows a pipe stake being used to form a radius or curve, whilst Figures 98 and 99 illustrate other types of tools and typical applications.

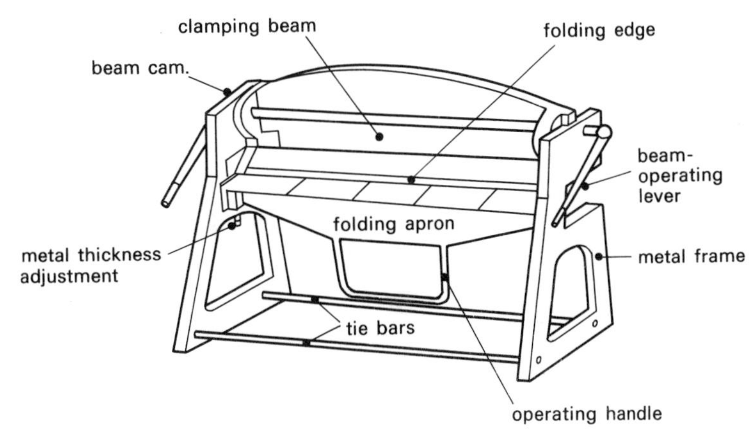

Figure 100 *Hand-operated folding machine*

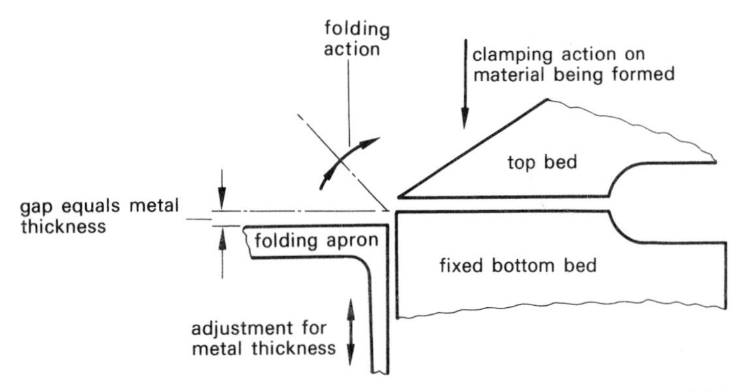

Figure 101 *Cross-section of part of a folding machine showing folding principles*

Sheet metal can be bent or folded using a number of different types of machine; one example of a hand-operated machine is shown in Figure 100, while the principle of operation of the machine is illustrated in a simple fashion in Figure 101.

Bending and folding allowances

Before cutting or forming sheet metal, the metal must be marked out accurately. If bending or folding operations are to be carried out, allowance must be made for the fact that material on the outer surface of the bend is stretched, whilst on the inner surface the material is compressed. At the centre of the thickness of the material there is a radius which neither extends nor compresses. This mean radius forms the basis for determining the bending or folding allowance.

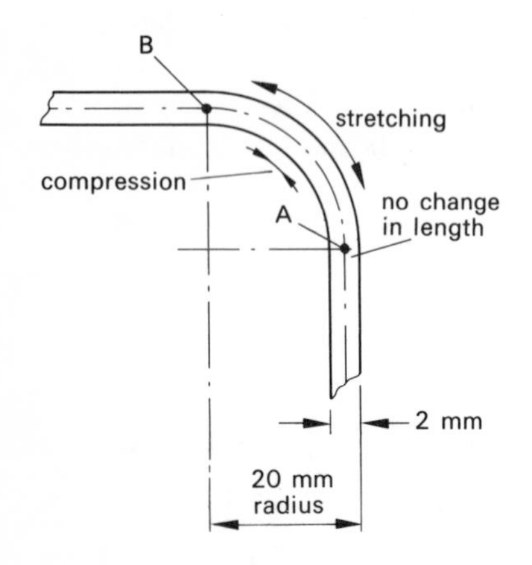

Figure 102 *Bending allowance*

Consider, for example, a sheet of material that is 2 mm thick; it is required to bend the sheet through 90° so that an outside radius of 20 mm is formed (Figure 102).

outside radius = 20 mm
material thickness = 2 mm
inside radius = 18 mm

$$\text{mean radius} = \frac{20 + 18}{2} = 19 \text{ mm}$$

The length of the circumference of the mean radius for a 90° bend (a quarter of circle) determines the bending allowance used in marking out.

circumference of a circle $= 2\pi r$

circumference of a quarter of a circle $= \frac{2\pi r}{4} = \frac{\pi r}{2}$

$\therefore$ length of mean radius (i.e. length of arc AB in Figure 102) $= \frac{\pi \times 19}{2} = 29.85$ mm

Example 2

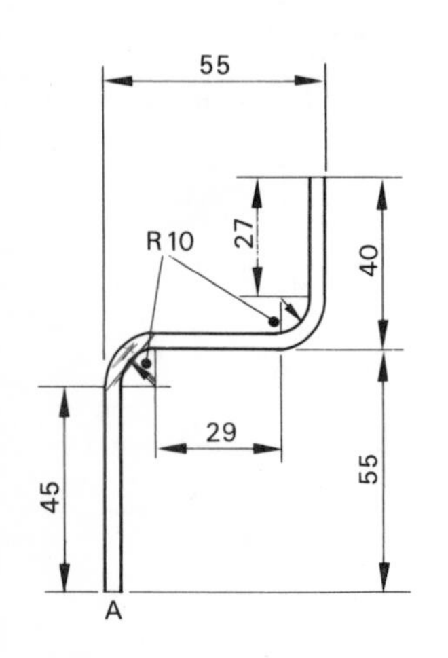

Figure 103 *Example 2*

Figure 103 shows the profile of a metal strip after bending. What length of strip should be cut to enable this profile to be generated?

Beginning at point A (see Figure 103), length of strip required

= 45 mm + bending allowance + 29 mm + bending allowance + 27 mm

The two bending allowances are the same because the radii of the two bends are the same.

$$\text{Bending allowance} = \frac{\pi r}{2} = \frac{\pi}{2}(10 + 1) \text{ mm} = 17.28 \text{ mm}$$

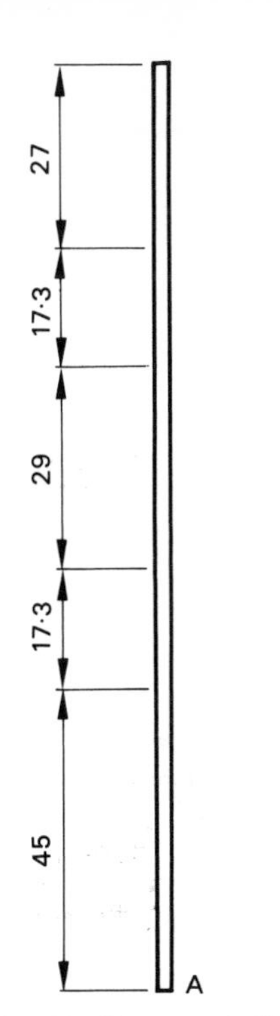

Figure 104 *Marking out before bending*

∴ length of strip to be cut

= (45 + 17.28 + 29 + 17.28 + 27) mm = 135.56 mm

The marked-out strip is shown in Figure 104.

The edges of items made from sheet metal are usually sharp, and may present a safety hazard. For this reason, and also to provide extra strength and rigidity, the edges can be folded. The simplest method of removing a sharp edge is to produce a *beaded* edge (Figure 105); the sequence of operations to form such an edge is shown in Figure 106.

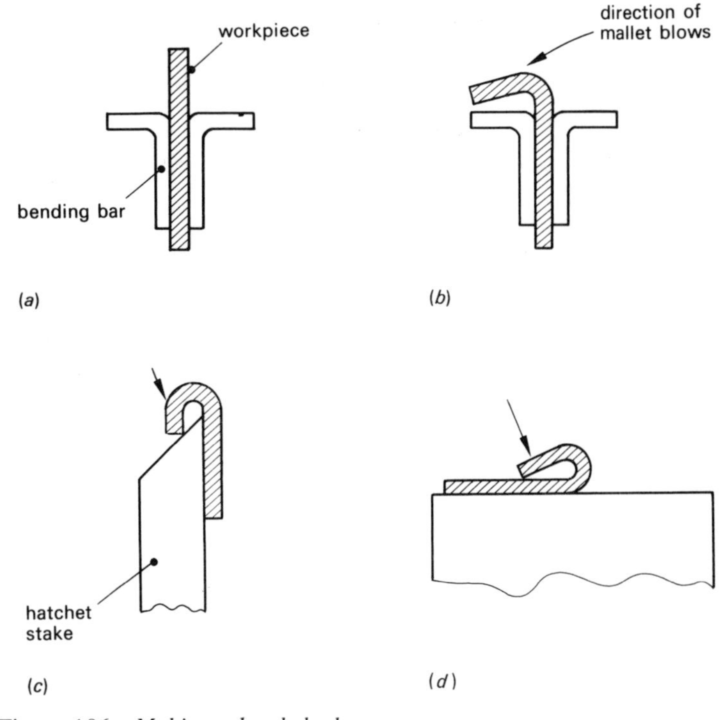

Figure 106 *Making a beaded edge*

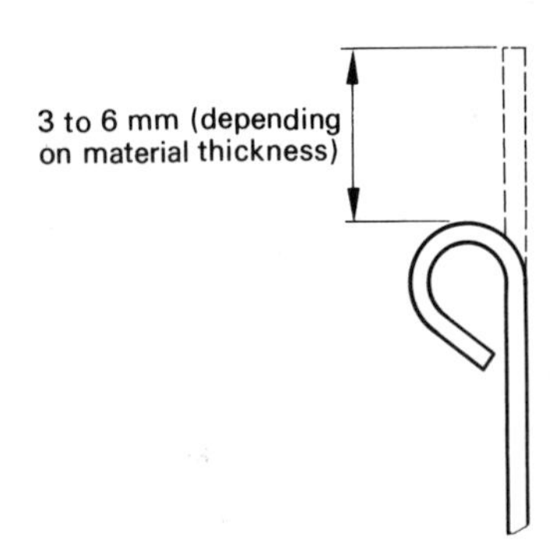

Figure 105 *Beaded edge*

Extra stiffness can be obtained by making a *wired* edge (see Figure 108). The method of manufacture is similar to that used to make a beaded edge, but a length of wire is included in the fold, so producing extra stiffness. Figure 107 illustrates the sequence for manufacturing a wired edge.

In order to illustrate some of the procedures described above, consider the manufacture of a small rectangular box.

The first step is to mark out the development of the box on sheet

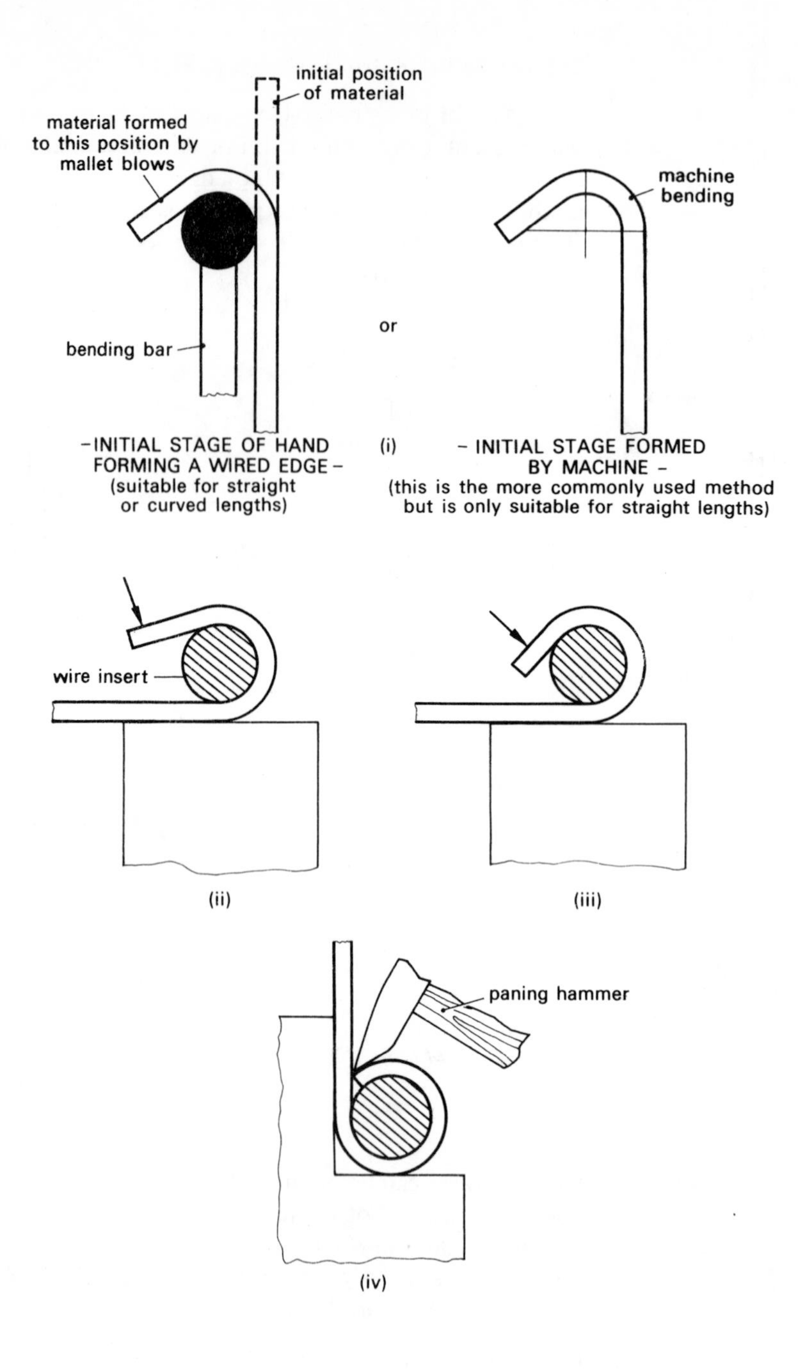

Figure 107 *Making a wired edge*

metal (Figure 109); bending allowances are included in the development. Flaps are also included in the development; the purpose of the flaps is to facilitate the assembly of the box once it has been bent into shape.

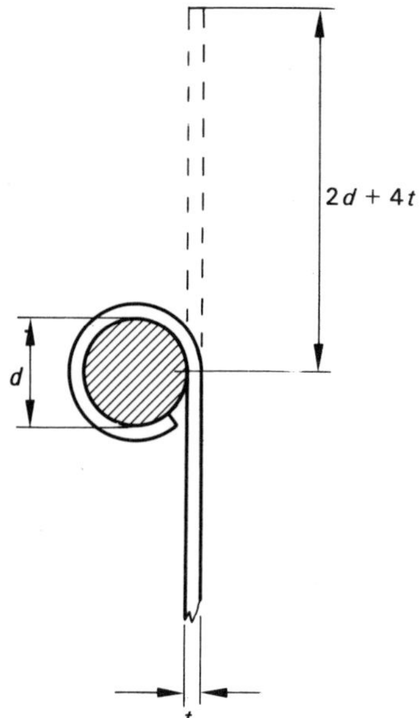

Figure 108 *Wired edge*

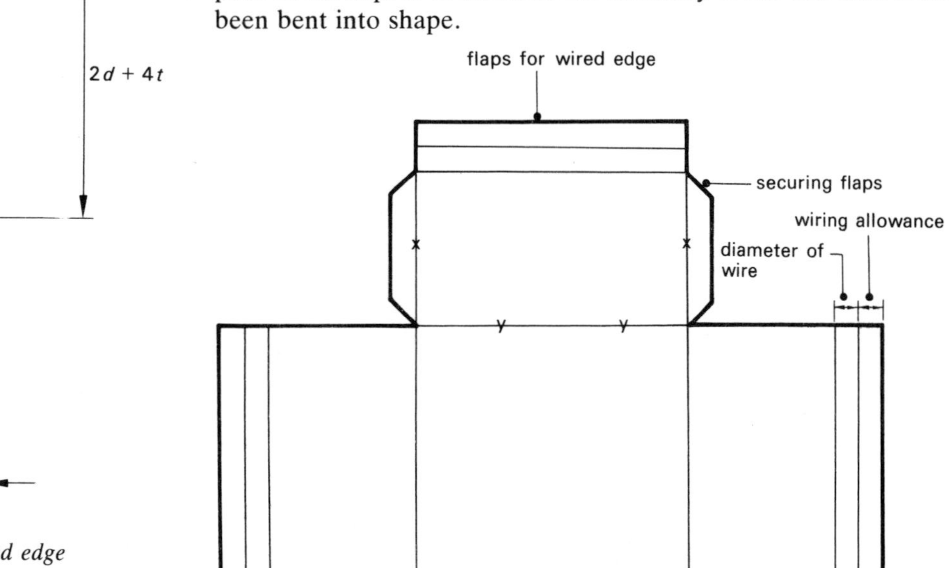

Figure 109 *Development of rectangular box*

The development can be cut out using snips, and then bent into the shape shown in Figure 110. If a wired edge is required, the wire can be arranged so that the joint in the wire is hidden under one of the folds, as shown in Figure 110. The box can then be completed as shown in Figure 111.

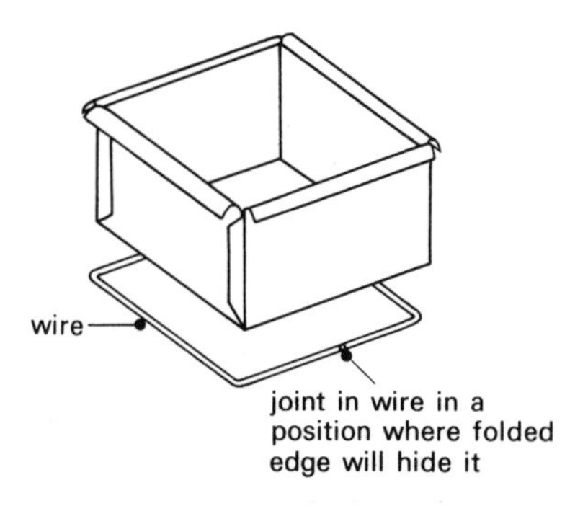

Figure 110 *Box with corners folded ready to receive wire*

Self-assessment questions

12 Name a cutting machine which can be used to cut the outside straight lines of Figure 109.

13 What type of handtool can be used to cut a circular arc in sheet metal?

14 Sketch a sequence for producing a beaded edge.

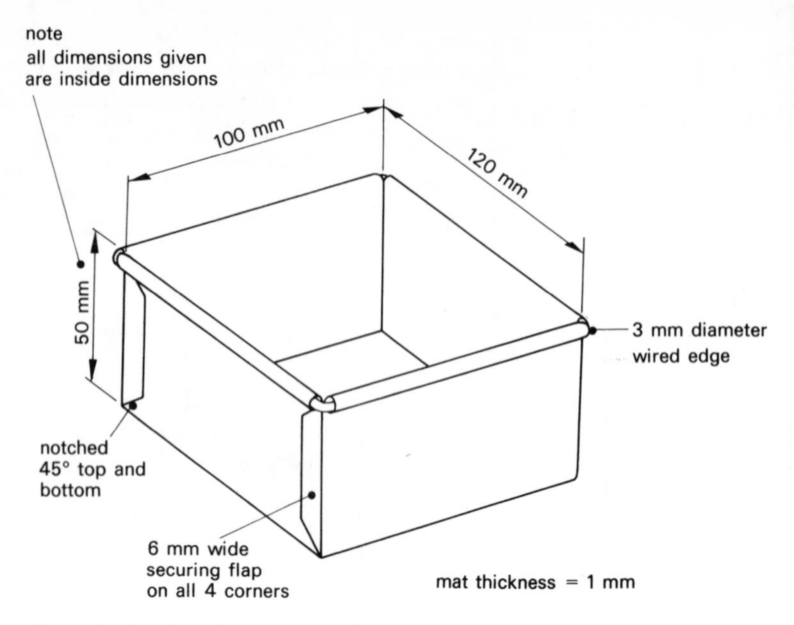

Figure 111 *Completed box with wired edge*

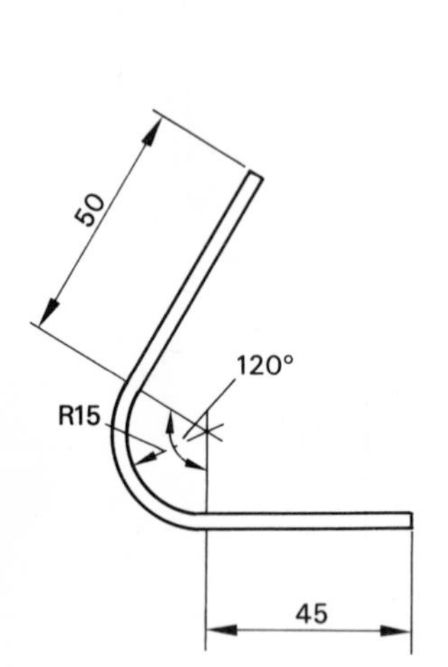

Figure 112 *Self-assessment question 16*

15 Give two reasons for providing folded edges on items made from sheet metal.

16 Figure 112 shows the cross-section of a small component to be made from sheet metal. Calculate the length of material required in order to make the component.

17 A wired edge is to be made using 4 mm diameter wire with 1 mm thick sheet metal. Study Figures 108 and 109 then calculate the allowance which should be made for the wired edge.

After reading the following material the reader shall:

7 Describe the function and uses of a centre lathe.

7.1 Identify the component parts and drive systems of a typical centre lathe, head stock, bed, slides, work-holding components.

7.2 Describe four methods of taper turning.

7.3 Describe methods of screw cutting using a single-point tool.

7.4 Identify the operations and procedure necessary to produce simple components on the centre lathe.

Solutions to self-assessment questions

12 A treadle guillotine is an appropriate machine.

13 Curved snips.

14 Figure 106 illustrates the sequence of operations.

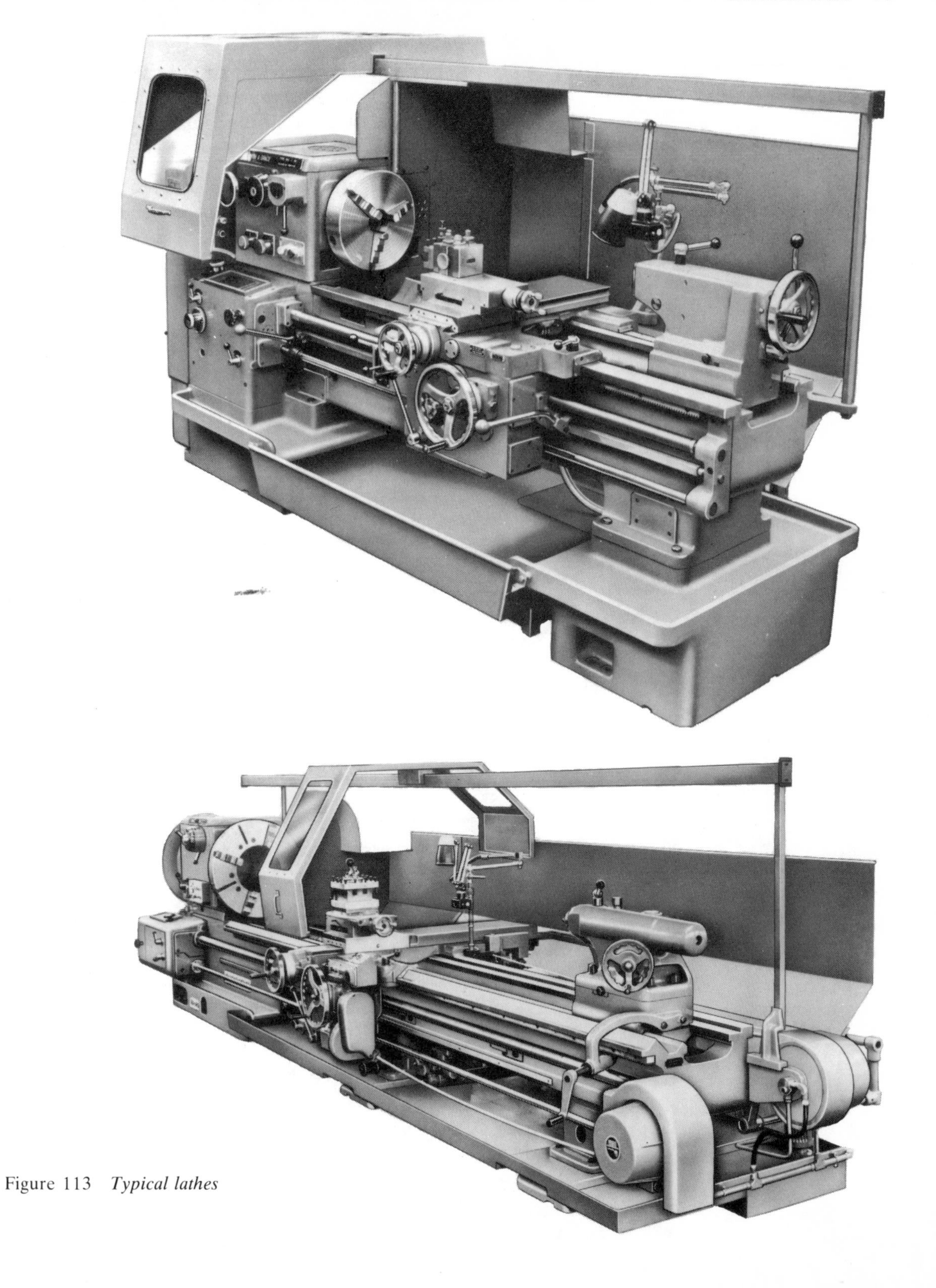

Figure 113 *Typical lathes*

Typical lathes are shown in Figure 113.

The centre lathe has probably the longest history of the machine tools that are found in a modern workshop. Early developments of the lathe were only suitable for shaping relatively soft materials such as wood or non-ferrous metals.

Following the development of the steam engine and greater understanding of materials, improved centre lathe design brought about greater flexibility in machining operations. Figure 114 shows an example of a centre lathe made at the beginning of the twentieth century; it can be seen that it has much in common with a modern centre lathe (Figure 115).

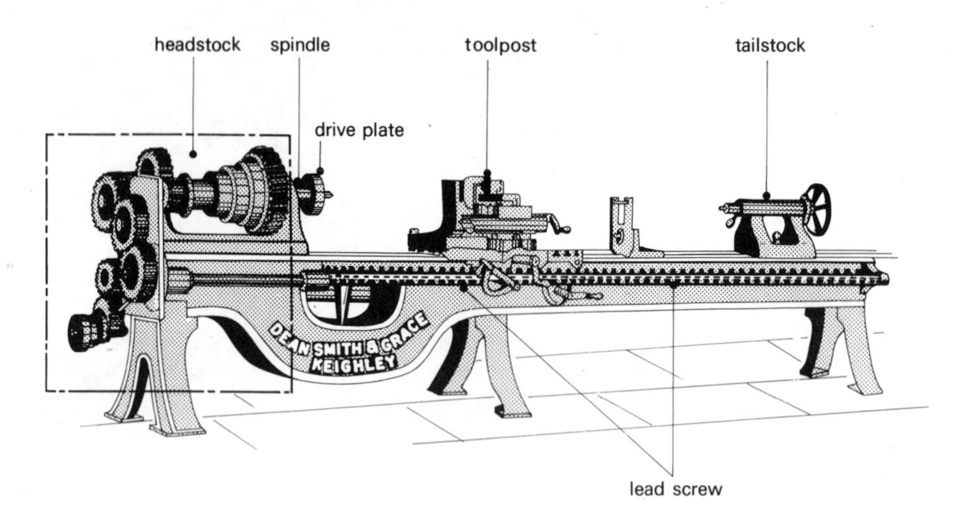

Figure 114 *Industrial lathe of about 1900*

Solutions to self-assessment questions

15 (i) The stiffness of the item is increased.

(ii) A folded edge is safer than a sharp edge.

16 Length of material = 50 mm + bending allowance + 45 mm

Inclusive angle of bend = 120°

Portion of circle occupied by the bend $= \frac{120°}{360°} = \frac{1}{3}$

Mean radius = (15 + 1) mm = 16 mm

Bending allowance $= \frac{2\pi r}{3} = \frac{2\pi \times 16}{3}$ mm = 33.5 mm

Length of material = (50 + 33.5 + 45) mm = 128.5 mm

17 Allowance for a wired edge = $2d + 4t$

where d = 4 mm and t = 1 mm

∴ Allowance = [(2 × 4) + (4 × 1)] mm = 12 mm

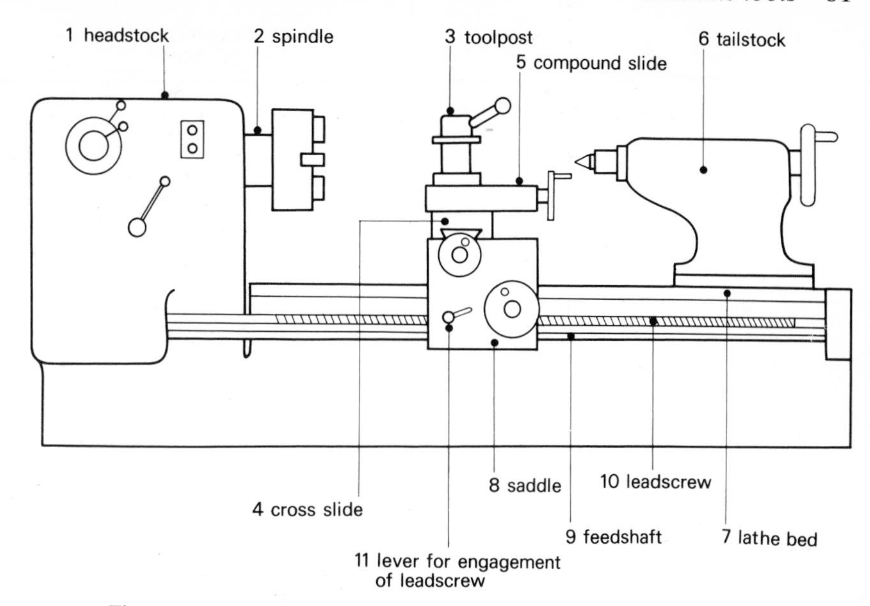

Figure 115 *Modern centre lathe*

The primary function of a centre lathe is to remove material to produce a cylindrically shaped workpiece. In producing a cylinder in the centre lathe, the workpiece is rotated (Figure 116), and a cutting tool is moved along the workpiece periphery, removing material as it travels. This combination of movements is termed *generating*.

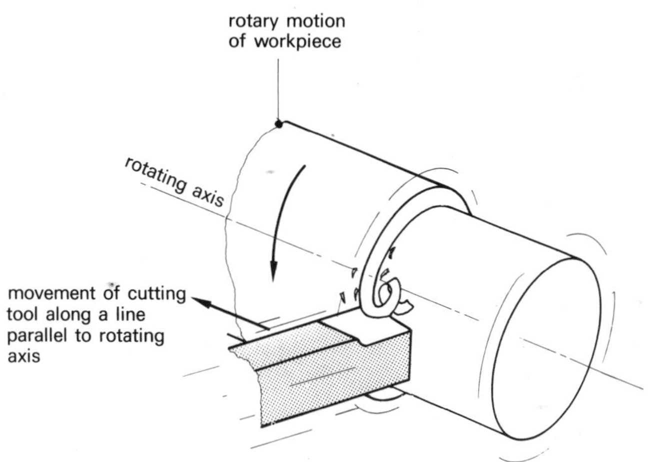

Figure 116 *Generating a cylinder*

Centre lathes can also be used to produce flat surfaces. Figure 117 shows how this may be accomplished. This operation is called *facing*; it produces a surface at 90° to the axis of the rotating workpiece.

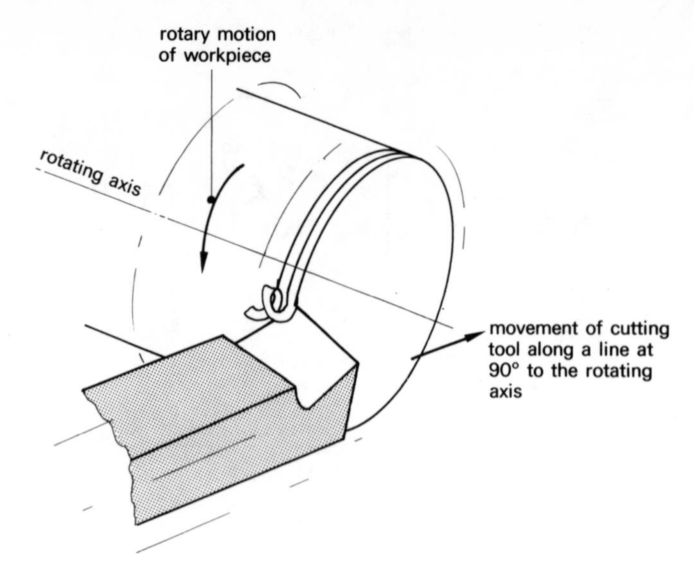

Figure 117 *Generating a flat face (facing)*

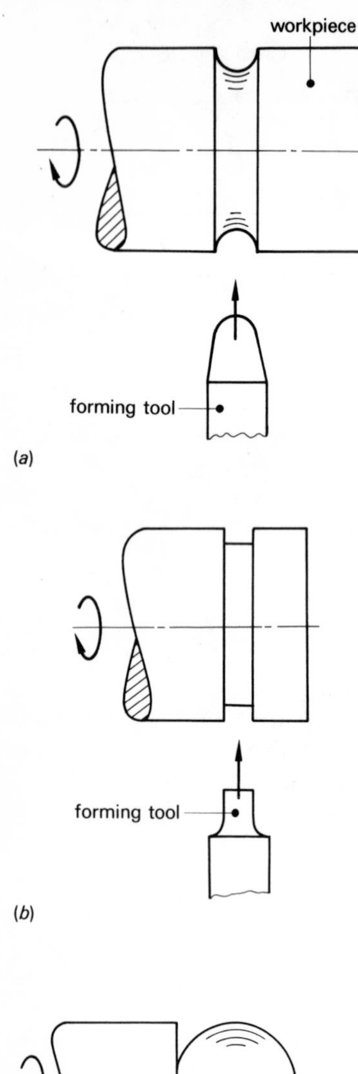

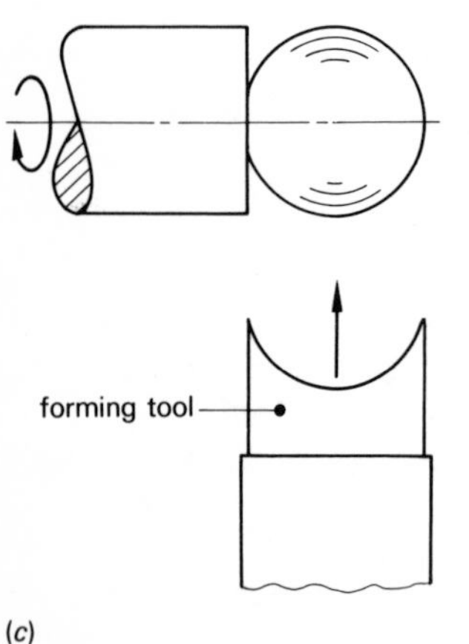

Figure 118 *Examples of forming*

If a cutting tool is fed into a rotating workpiece, the shape produced in the workpiece resembles the profile of the cutting tool. By grinding the cutting tool to a particular shape, that shape can be reproduced in a workpiece. This means that components can be produced quickly and accurately even though the diameters at various points are different. The process is called *forming* and is illustrated in Figure 118.

To operate a centre lathe effectively, it is important to understand the functions of the main parts of the machine tool.

The headstock

In order to machine different materials or to produce different surface finishes, the lathe must be capable of rotating the workpiece at different speeds. The most common method of varying rotational speeds involves the use of a gear train; that in the centre lathe is housed in the headstock.

Transmission gears occupy the space available inside the headstock casting and provide the drive to the spindle, feedshaft and leadscrew (items 2, 9 and 10 in Figure 115). The hollow box-like casting gives rigidity, an important characteristic on any machine tool. On the outside of the casting are located the various operating controls shown in Figure 119.

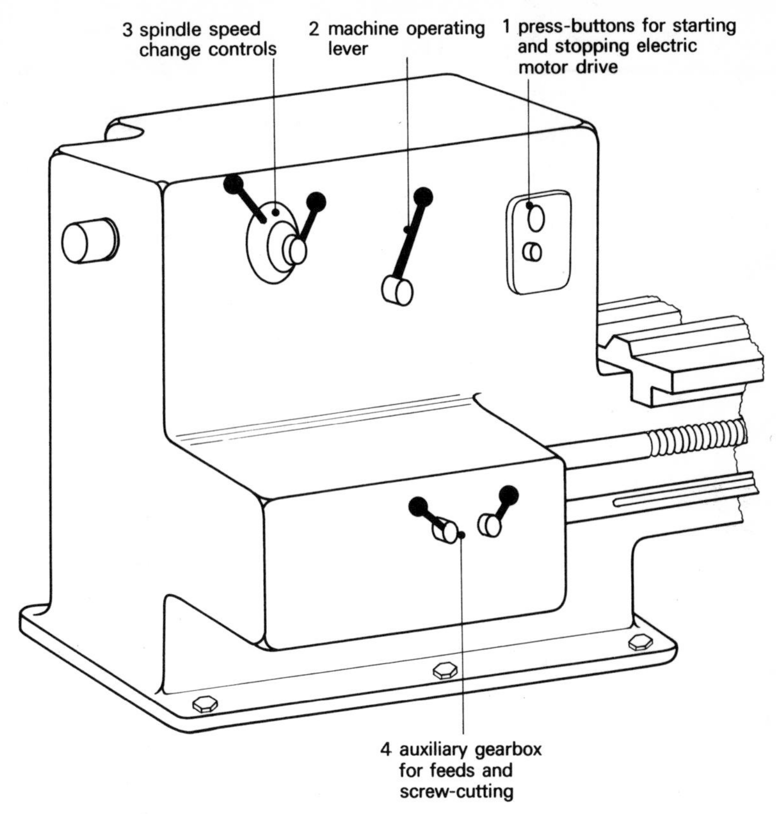

Figure 119 *The headstock*

The lathe bed and slideways

To generate a cylinder it is essential that the cutting tool is maintained at a fixed distance from the rotating axis, so giving a constant radius as shown in Figure 120. It is important also to adjust the height of the cutting tool so that it is neither above nor below the rotating axis.

The *bed* of a centre lathe (item 7 in Figure 115), is the 'backbone' of the machine. It has two main functions:

(i) To give the necessary stiffness to resist the forces and resulting deflections that occur during cutting operations.

(ii) To ensure that the cutting tool moves accurately along a path that is parallel to the machine spindle axis.

The centre lathe bed is a box-type casting made of grey cast iron. The use of a casting enables the complicated shape to be produced relatively easily, with extra strength and stiffness being provided by *ribs* (Figure 121). The requirements of a lathe bed are:

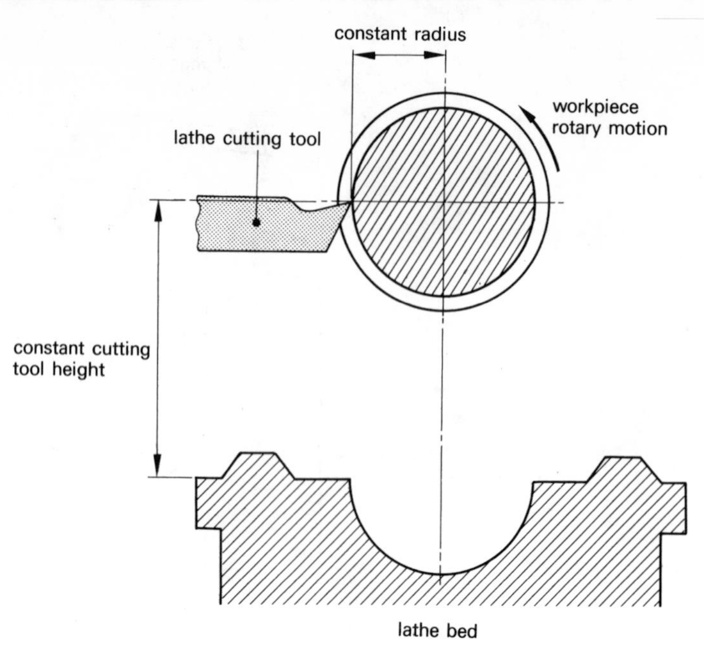

Figure 120 *Requirements to produce a cylinder on a centre lathe*

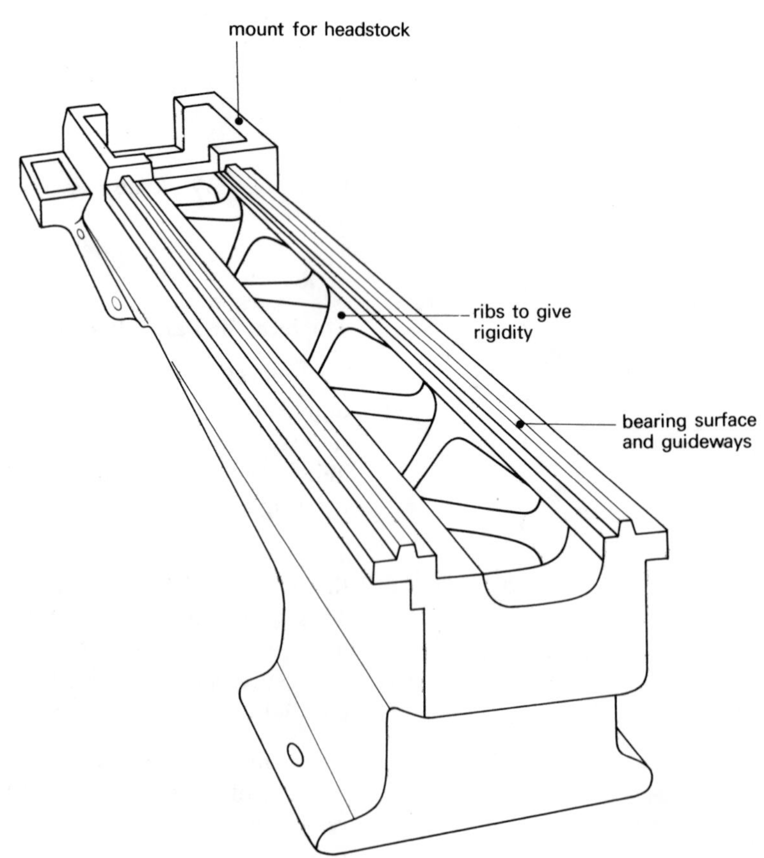

Figure 121 *Lathe bed casting*

(i) A good, even wear resistance to ensure long useful life
(ii) Low frictional resistance between moving parts
(iii) The damping of any vibrations set up in the machine during cutting operations

Grey cast iron fulfils these requirements to a large extent, and so it is used extensively for lathe bed castings.

By providing the tailstock (item 6 in Figure 115) and the saddle (item 8 in Figure 115) with separate bearing surfaces, some problems of wear are reduced (Figure 122).

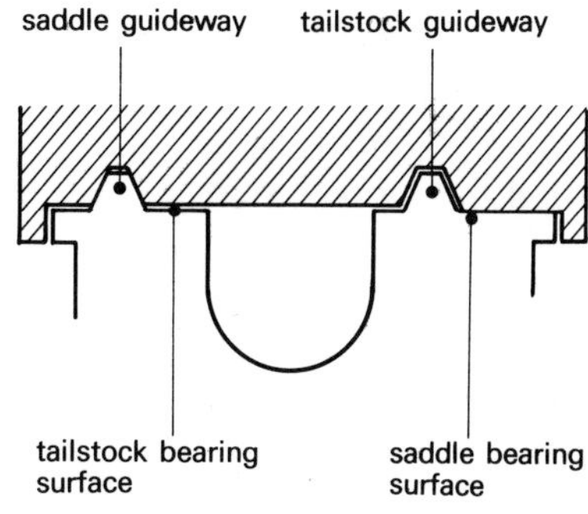

Figure 122 *Section through a lathe bed*

The saddle

The saddle fits directly on to the lathe bed, as shown in Figure 123.

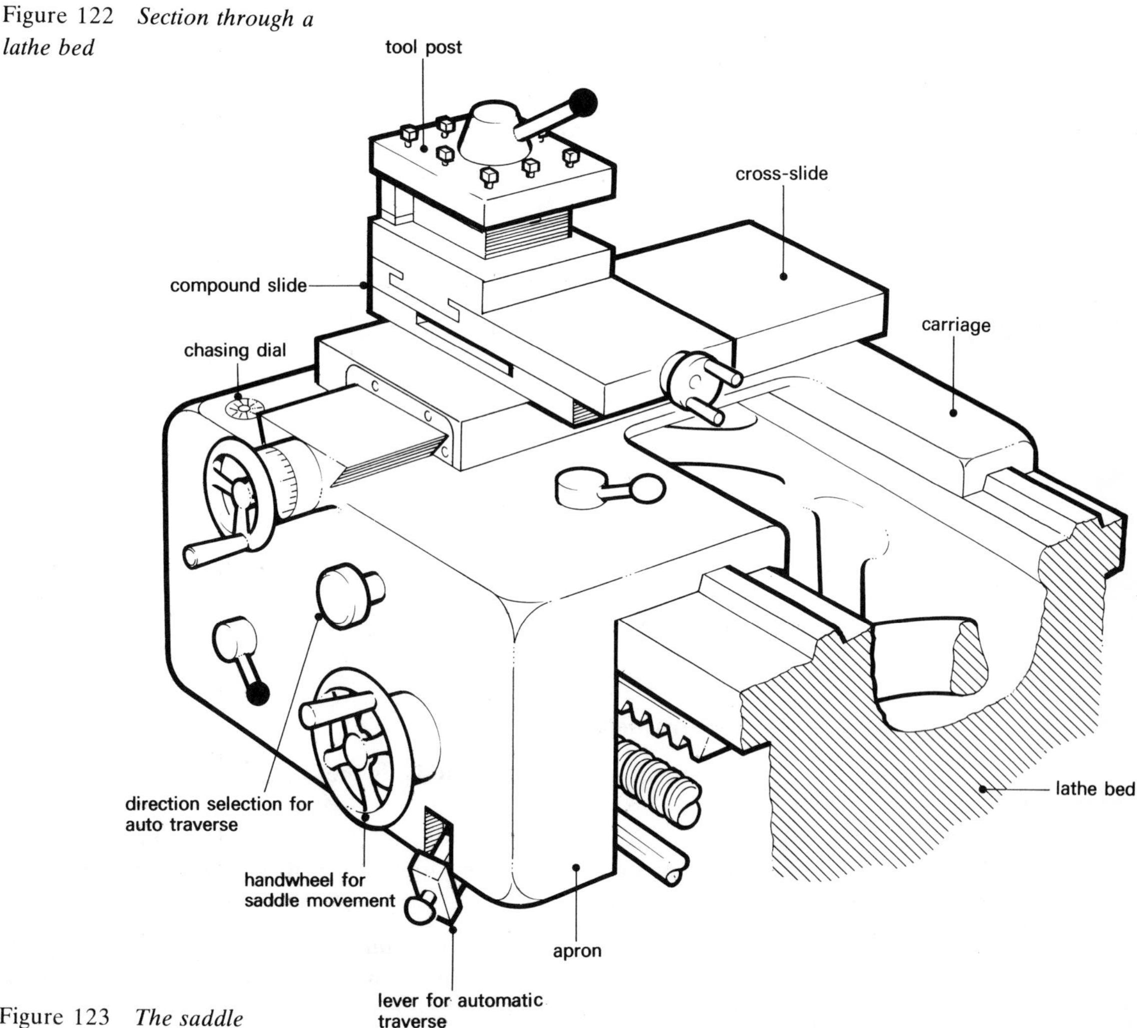

Figure 123 *The saddle*

Its purpose is to provide movement that is parallel to or perpendicular to the spindle axis. Two parts make up the saddle: the *apron*, which carries the controls, and the *carriage*, which lies across the lathe bed.

Movement of the saddle along the lathe bed is carried out in one of two ways:

(i) By manual operation of a hand wheel (Figure 123), or
(ii) By an automatic traverse that is fed by the feedshaft (item 9 in Figure 115).

It is this movement that provides the motion of the cutting tool when generating a cylinder.

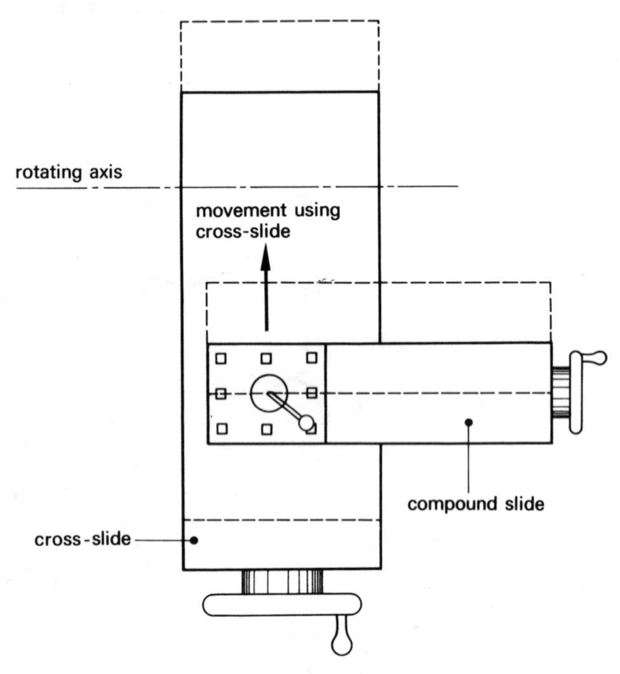

Figure 124 *Cross-slide motion*

The cross-slide

Mounted on top of the saddle is the *cross-slide* (item 4 in Figure 115). This provides:

(i) Support for the compound slide (item 5 in Figure 115)
(ii) Movement of the cutting tool at 90° to the rotating axis of the workpiece (Figure 124)

Motion of the cross-slide is obtained by a hand wheel, but on the majority of centre lathes a gear mechanism is fitted to give automatic traverse both towards the spindle axis and away from it.

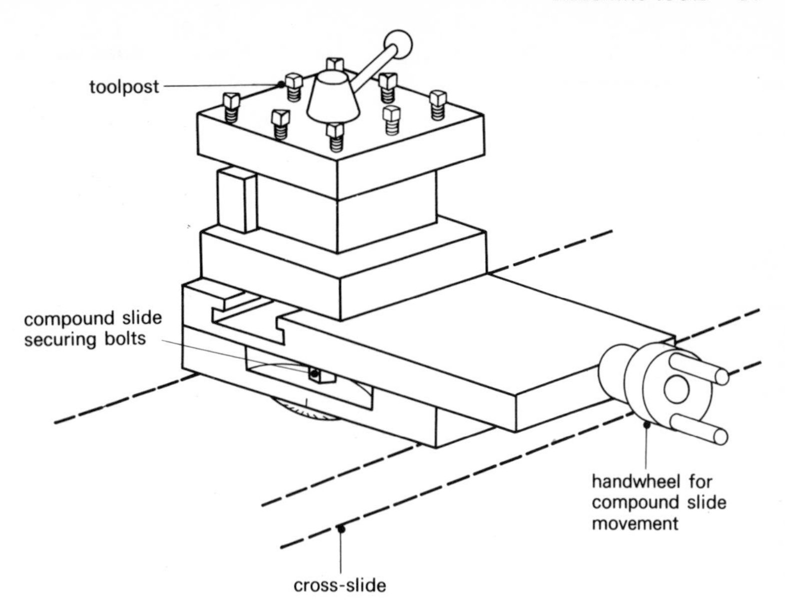

Figure 125 *Compound slide*

The compound slide

The compound slide (item 5 in Figure 115) is secured to the cross-slide and has two main functions:

(i) To provide location and support for the toolpost (item 3 in Figure 115).

(ii) To enable the tool to be placed at an angle to the axis of the spindle.

Compound slide movement is obtained by using a hand wheel; no automatic traverse is provided. Two bolts secure the compound slide base, as shown in Figure 125. If these are slackened, the slide can be rotated about its mounting. A circular scale on the base indicates the angle through which the slide has turned.

The toolpost

Many varieties of toolpost are available, each carrying out two main functions:

(i) Positioning the cutting tool

(ii) Securing the cutting tool in that position

To enable the cutting action to be carried out efficiently and produce the desired surface finish, it is important that the tool tip is set to the correct height. This is on the spindle axis. Using a centre (set in the tailstock – Figure 126), the correct height of the tool tip is judged. Thin metal strips are used to 'pack' the tool to the correct height. This procedure requires considerable skill.

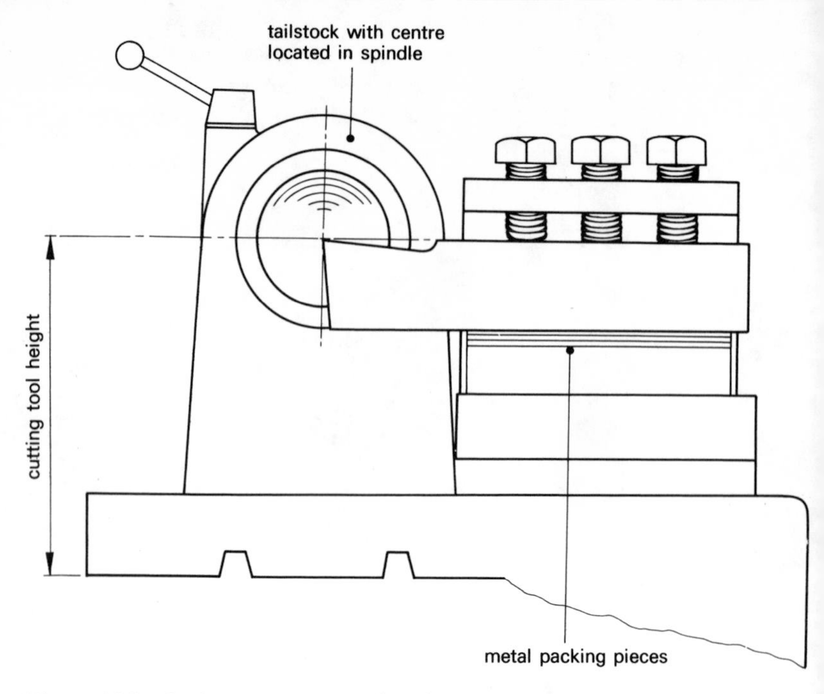

Figure 126 *Packing a cutting tool to the correct height*

The toolpost shown as an exploded view in Figure 127 is called a *single toolpost*. Such a toolpost can accommodate only one cutting tool at a time. It may prove inconvenient when several cutting tools are required in machining a component. In such a case a *four-way toolpost* can be used (Figure 128).

The toolpost is rotated around a central spindle to allow each of any of four tools to be positioned as required.

The tailstock

The tailstock is shown in cross-section in Figure 129.

It is positioned at the end of the lathe bed opposite to the headstock. The tailstock body is produced in cast iron and is machined on its base to enable it to slide along the guideways and bearing surfaces of the lathe bed.

The barrel moves within the body. This movement is actuated by a handwheel attached to a square threaded screw. The top surface of the barrel is graduated so that adjustments may be made without the use of a rule. The graduations become visible as the barrel extends out of the body.

At the end of the barrel opposite to the handwheel is a tapered hole. This has a morse taper which matches the morse tapers on centres,

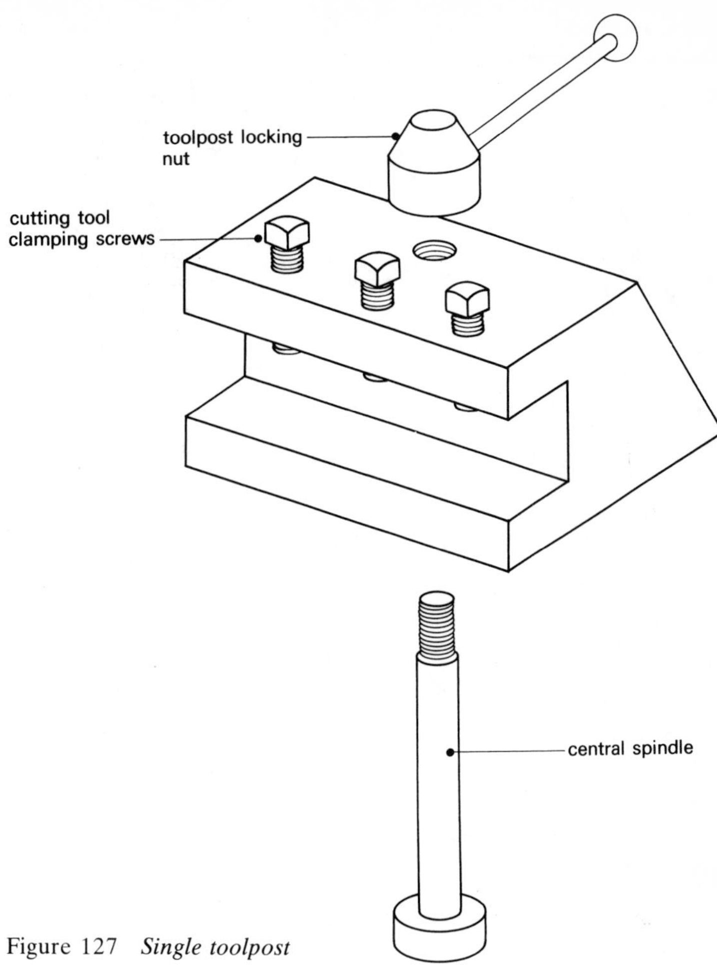

Figure 127 *Single toolpost*

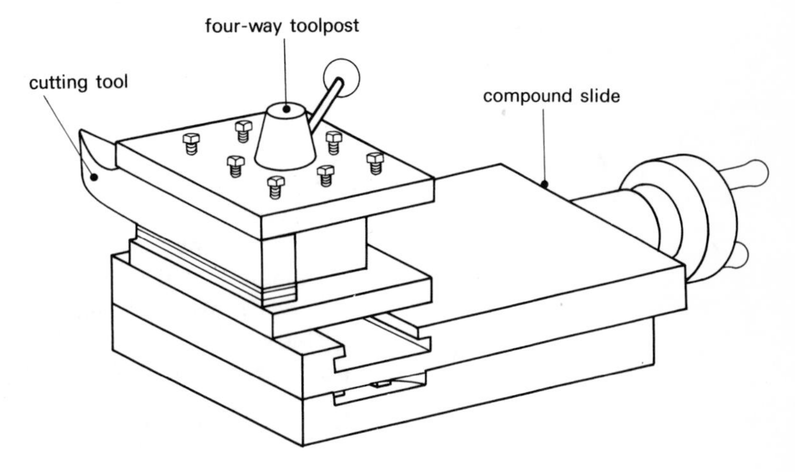

Figure 128 *Four-way toolpost*

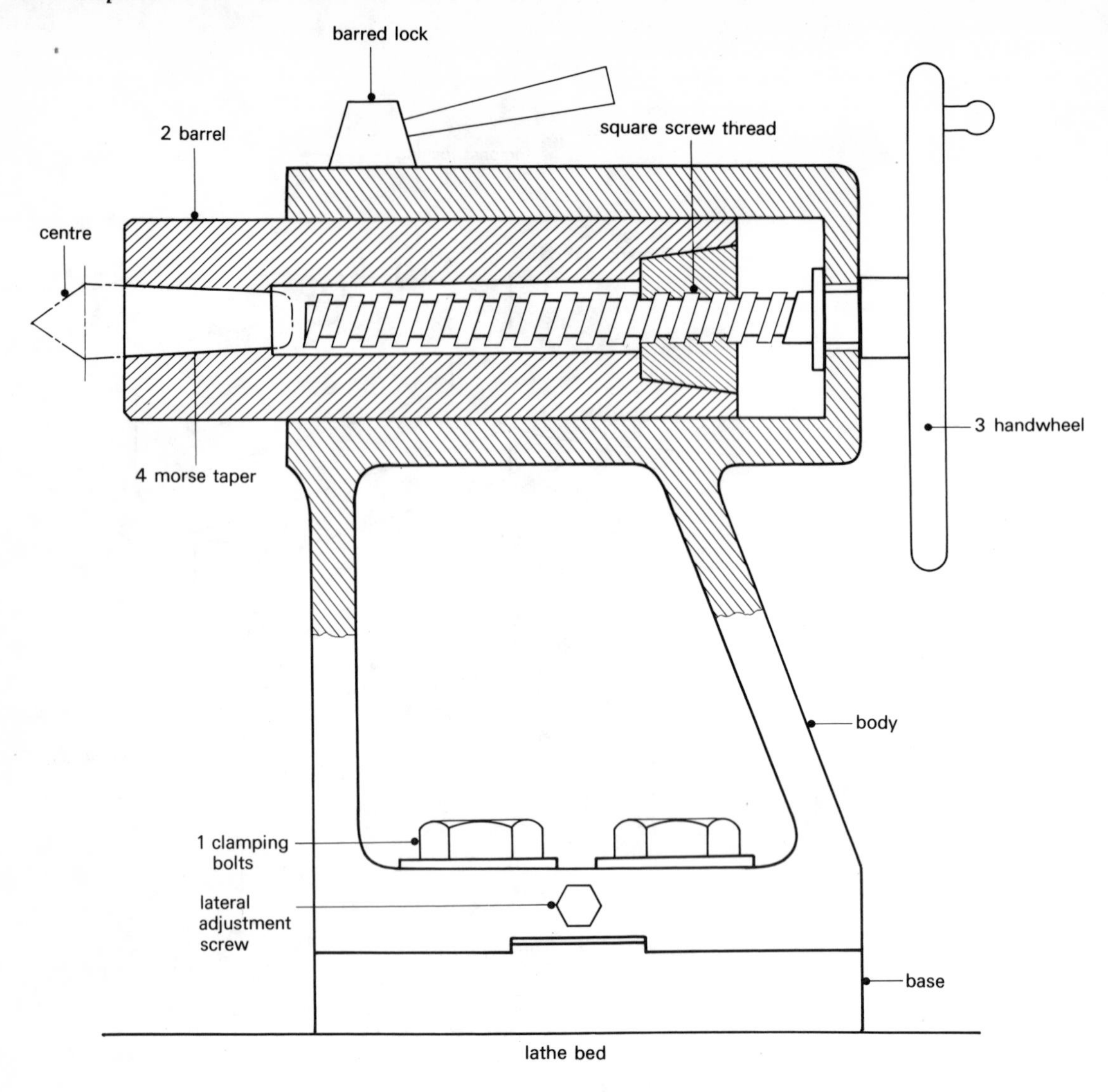

Figure 129 *Section through a tailstock*

drills, reamers, etc. When fitted, the centre (or the tools) should be in line and on centre with the centre of the headstock spindle. The centre (and tools) are positioned by hand, a fit occurring as the tapers mate and grip each other. Removal of the centre or tool is obtained by turning the handwheel so that the barrel enters as far as possible inside the body. The end of the screw thread makes contact with the centre or tool, pushing it from the barrel.

The barrel can be locked in any position by the barrel lock. Similarly the tailstock body can be clamped to the bed by means of a lever on the side of the body. Sideways adjustments can be made by using the lateral adjustment screw. This moves the centre of the barrel off centre with the headstock spindle and is used for machining slow tapers (a method described later).

The height from the lathe bed to the barrel axis is important and is a fixed dimension. The barrel must be aligned with the nose spindle as shown in Figure 130.

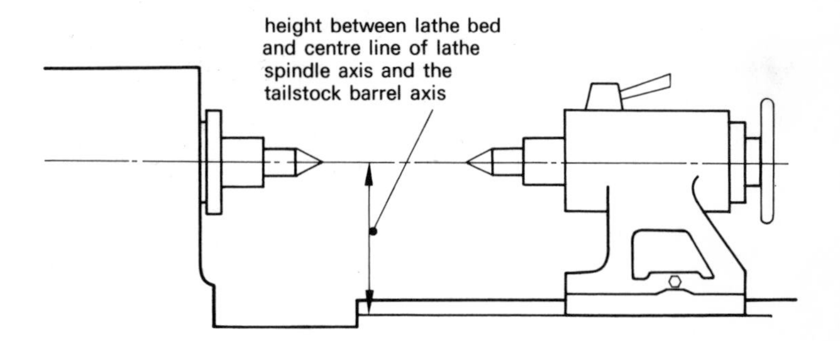

Figure 130 *Alignment of the tailstock barrel*

Work-holding devices

When machining a component on a centre lathe the material must be held in some device. This device:

(i) Locates the work in the desired position.
(ii) Secures the work in that position during the machining.
(iii) Enables the work to be rotated, using the energy available at the spindle which protrudes from the headstock.

The devices most commonly used for holding work on the centre lathe are:

(i) Driving plate and centres
(ii) Three-jaw chuck
(iii) Four-jaw chuck
(iv) Faceplate

(i) *Driving plate and centres*

This method was used in the early days of the development of the centre lathe. It is from this type of arrangement that the centre lathe derives its name.

The driving plate is located and locked to the nose of the spindle. Centres are fitted:

(a) into the end of the headstock spindle, and
(b) into the tailstock barrel.

The material from which the component is to be machined is

centre-drilled at each end to provide a location for the centres and a lathe carrier or dog is fitted to one end. The workpiece is then placed between the centres and locked in position with the carrier locating the driving pin, which is attached to the driving plate (Figure 131). A typical dog or carrier is shown in Figure 132.

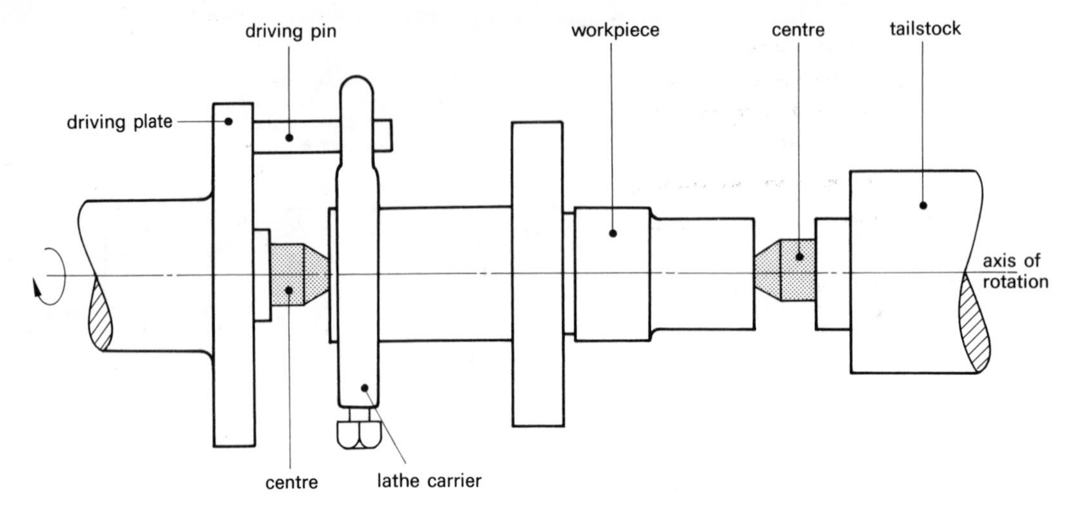

Figure 131 *Driving work between centres*

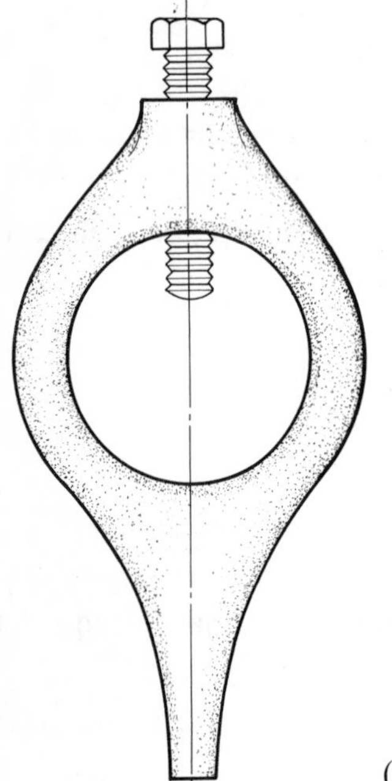

Figure 132 *Driving dog or carrier*

The centres usually have a 60° included angle point, and the shank has a morse taper, which fit the tapers in both the headstock spindle and the tailstock barrel. A typical centre is shown in Figure 133.

When the work is positioned between the centres, the tailstock should be positioned and locked so that the centres just grip the material. There must be no axial movement.

Provided that the centre in the headstock spindle runs true and is in line with the tailstock centre, a component may be removed from the lathe at any stage (for additional marking-out, measuring and other operations), and then replaced between the centres to complete the machining processes. An advantage of this method of turning is that heavy cuts can be taken when a good deal of material must be removed. The process is particularly suitable for machining items such as shafts (either short or long).

One disadvantage is that the holes in which the centres locate remain in the component unless an additional operation is carried out, but in many cases the small holes are acceptable.

(ii) *Three-jaw chuck*

A 3-jaw chuck is shown in Figure 134.

The device provides self-centring location. All the jaws move an

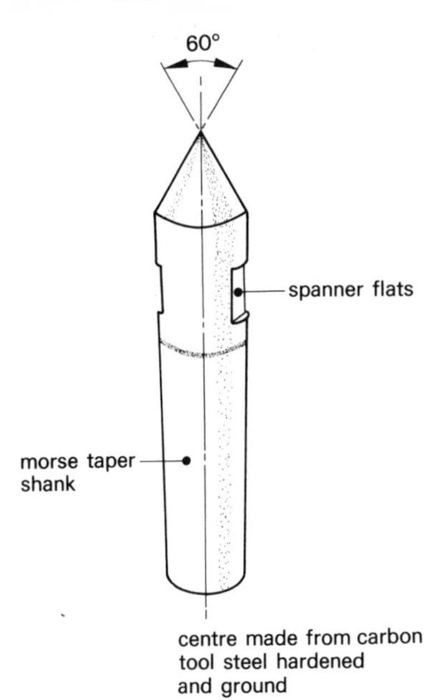

Figure 133 *Lathe centre*

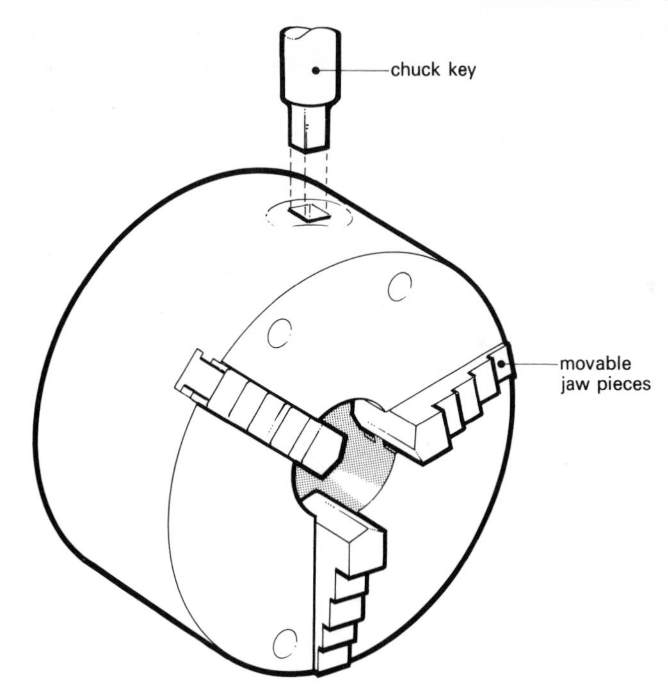

Figure 134 *Three-jaw self-centring chuck*

equal amount when adjustments are made by the chuck key at any of the square sockets.

The chuck is fitted to the spindle nose. It is suitable for holding cylindrical work or work having a number of sides which are divisible by three. It is advisable to use a three-jaw chuck for components which protrude only a short amount from the chuck. Large unsupported lengths are likely to distort. When the workpiece must be held in a chuck, additional support can sometimes be provided by a centre in the tailstock.

Light cuts must be used when using these chucks because the work may slip in the jaws. During use, the chucks wear, with the result that work may become less accurate. It is advisable therefore that machining operations be completed without changing the set-up. Re-assembled work is unlikely to be as accurate, concentric and repeatable as work carried between centres.

(iii) *Four-jaw chuck*

This chuck is much heavier than the three-jaw chuck, but it fits on the spindle nose in the same manner as the three-jaw chuck. Each jaw is individually adjusted and moves along its own slot. Because it

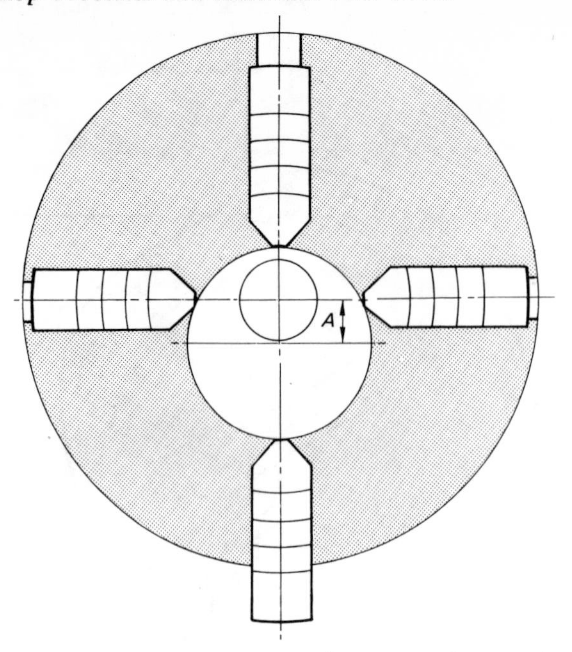

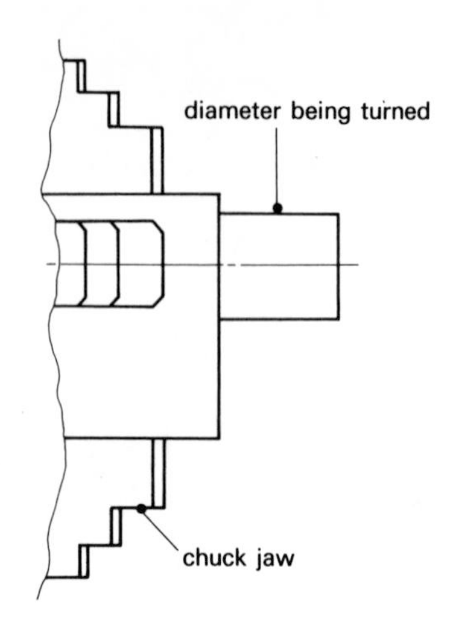

Figure 135 *Turning eccentric centres*

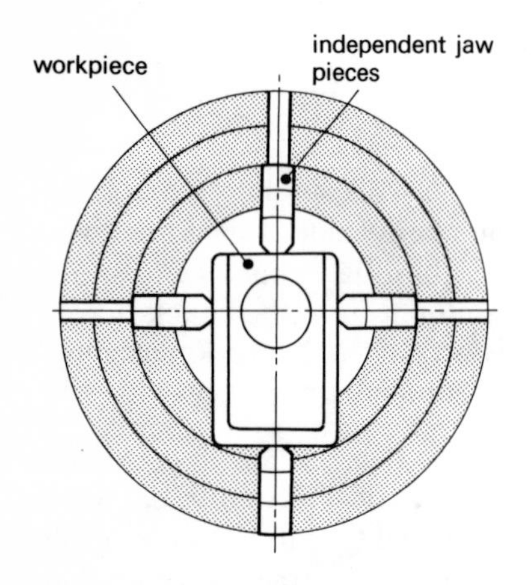

Figure 136 *Four-jaw chuck*

is designed differently, this chuck grips the work much more tightly than a three-jaw chuck.

One advantage of the four-jaw chuck is that work can be located in the centre to run true or off centre. An example of off-centre machining is shown in Figure 135.

Perhaps one of the most useful applications of the four-jaw chuck is to hold square or rectangular material positioned either centrally or off centre. An example is shown in Figure 136.

When using black bar (i.e. hot rolled material), the material should be machined in a four-jaw chuck.

Since each jaw is positioned individually, the setting time is greatly increased, when compared with the setting time of a three-jaw, but for highly accurate work the effort is justified.

(iv) *Faceplate*

Typical examples of a faceplate are shown in Figure 137.

Faceplates are basically flat discs containing slots. The faceplates are located and secured on the spindle nose. The surface provided by the face of the faceplate enables a component which is irregular in shape to be clamped to it so that part of the component can be machined. Castings or fabricated assemblies are frequently machined in this way.

Since irregular-shaped components are clamped to the faceplate an 'out-of-balance' force is frequently created when the component rotates. To counteract this out-of-balance effect, a counterweight is used. An example of this is shown in Figure 138. Frequently more than one counterweight is needed.

In general, most of the work machined using a faceplate is work which cannot be held in a chuck. Sometimes the faceplate may be used to finish work already commenced in a chuck but where the chuck is unsuitable for the final operations.

Figure 137 *Two typical faceplates for use on a centre lathe*

Taper turning

Tapers can be turned in a number of ways on the centre lathe. The most common methods are:

(i) Using a form tool
(ii) Using the compound slide rotated at an angle to the workpiece axis
(iii) Offsetting the tailstock

(i) *Taper turning using a form tool*

Tapers produced by means of a form tool can only be machined if the length of the tapered portion is short. A cutting tool is ground at the appropriate angle and positioned in the tool post (Figure 139). The work, held in a chuck or on a faceplate, is rotated and the tool fed into the work at right angles to the axis of the rotation. One problem with this method of taper turning is the possibility of vibration and chatter due to the length of the cutting tool in contact with the workpiece. This may result in a poor finish.

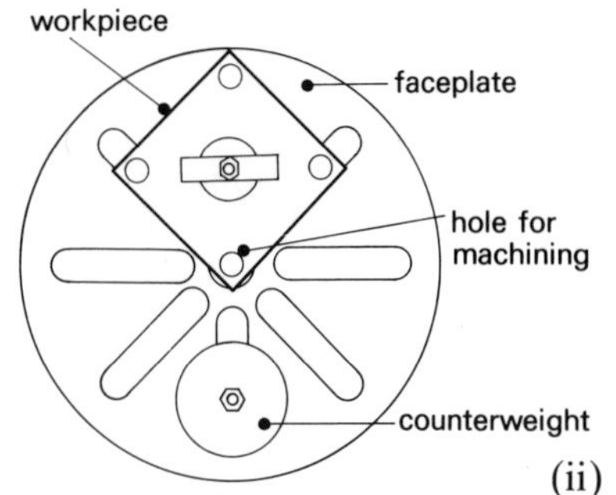

Figure 138 *Workpiece set on a faceplate*

(ii) *Taper turning by use of compound slide*

The compound slide (on which the toolpost is mounted) is usually set parallel to the axis of rotation when turning cylinders. If the slide is positioned at an angle to the axis of rotation, a taper can be turned on the workpiece. The base of the slide is graduated in degrees and if this is turned to half the included angle of the taper, the appropriate taper can be produced (Figure 140).

As the work is rotated, the cutting tool is moved at an angle to the axis of rotation by turning the handwheel on the end of the compound slide. This motion is not power driven. The handfeed used can vary, resulting in surface variations. The work also becomes tedious and causes operator fatigue. The length of the taper is limited to the length of travel of the compound slide. Both internal and external tapers may be produced.

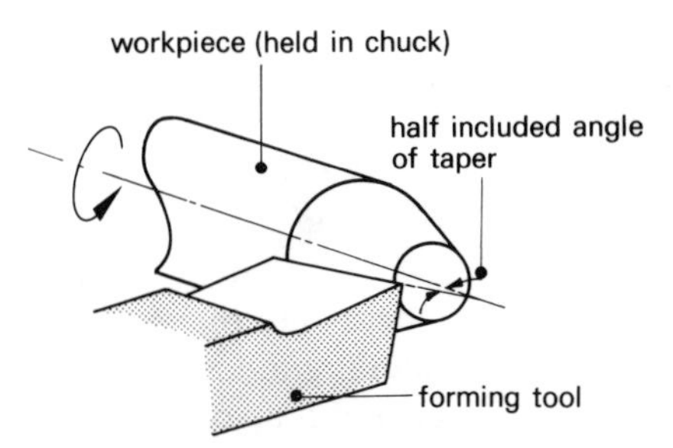

Figure 139 *Taper turning using a form tool*

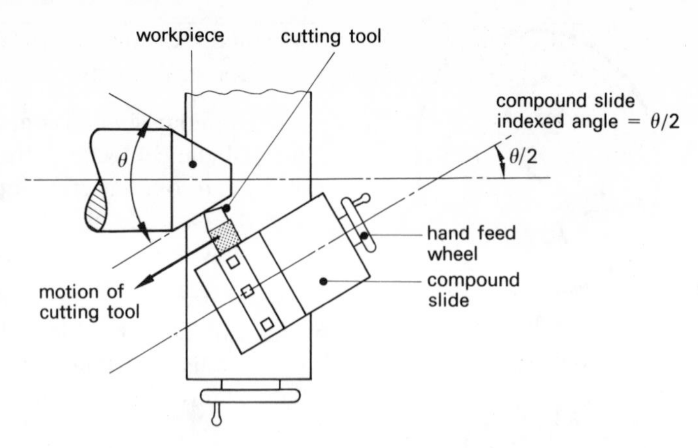

Figure 140 *Taper turning using the compound slide*

(iii) *Taper turning with the tailstock offset*

This method can only be used to produce a slow taper (small included angle of taper) and can only be used when machining between centres. The whole of the tailstock is moved sideways by adjusting the lateral screw so that the axis of the barrel is now offset and parallel to the lathe bed. This means that the axis of the workpiece is inclined at an angle to the spindle axis. See Figure 141.

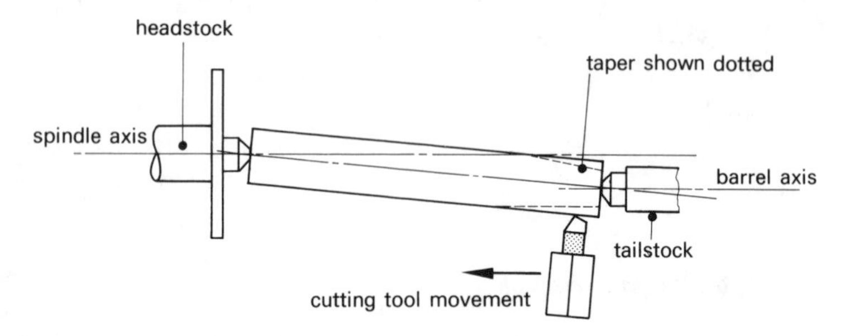

Figure 141 *Cutting a taper by offsetting the tailstock*

The work is driven by the peg on the driving plate. As the work rotates, the cutting tool moves parallel to the lathe bed (as it does when cutting cylinders). Since the material is inclined at an angle, the tool produces a taper on the workpiece. The amount of offset needs to be calculated each time because the lengths and angles vary for every component. The tailstock has no lateral graduations, so each offset must be accurately measured by the machine operator.

After machining is completed, the tailstock must be returned to its correct position. The alignment of axes must be checked or inaccuracies may be encountered when further machining operations are carried out.

Figure 142 *Taper-turning attachment*

Taper-turning attachment

The taper-turning attachment is fitted on the rear of the bed. It is commonly used on modern lathes. As the saddle moves along the bed, the attachment controls the cutting tool movement (Figure 142). The process is suitable for tapers with included angles up to 30°.

Screw cutting on the lathe

When a screw is to be cut using a centre lathe, the tool must move along the bed at a precise rate in relation to the rotation of the work. The tool movement is controlled by the leadscrew, which is driven from the main gearbox through a train of gears. The leadscrew passes through the apron. A lever on the front of the saddle apron is moved to engage a split phosphor-bronze nut which then drives the saddle along the bed at a predetermined rate. The lever is shown in Figure 115, page 81.

The tool must be ground to the correct shape and form. Usually there is no top rake on a tool used for screw cutting. When the correct height of the cutting point has been set, the tool must be set square to the work. This is done by using a screw-cutting and centre plate gauge as shown in Figure 143.

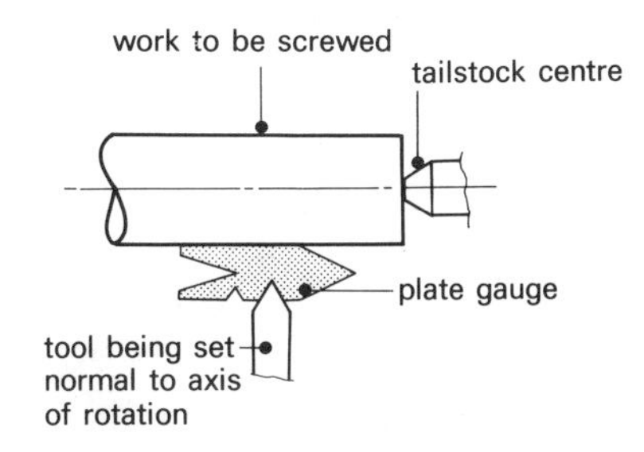

Figure 143 *Screw-cutting and centre gauge plate in use*

After setting the tool to the correct height and correct position the appropriate feed rate is selected by setting the levers on the auxiliary gear box at the base of the headstock. In addition, the machine operator must choose the correct rotational speed for the material to be cut.

There are two commonly used methods of screw cutting. Both involve a number of passes by the tool to obtain the full depth of the thread.

The first method is to set the tool to the outside diameter of the work and feed in normal to the axis of rotation. (A disadvantage of this method is that the tool cuts on both sides of the vee form; this does not always provide a satisfactory finish.) When the first cut has been taken, a helical groove is seen on the workpiece. The cutting tool is returned to the position where it began to cut the thread. It is now important for the tool to engage the same groove for each cut that is taken. One method of doing this is to use the chasing dial (Figure 123, page 85). The dial has a number of markings which are used for locating the correct position on the leadscrew. This ensures that the tool cuts in the same groove for each cut. Successive cuts are taken until the required depth is obtained.

A second method of screw cutting, and one more widely used, is to set the compound slide to an angle equal to half the angle of the thread. This is shown diagrammatically in Figure 144.

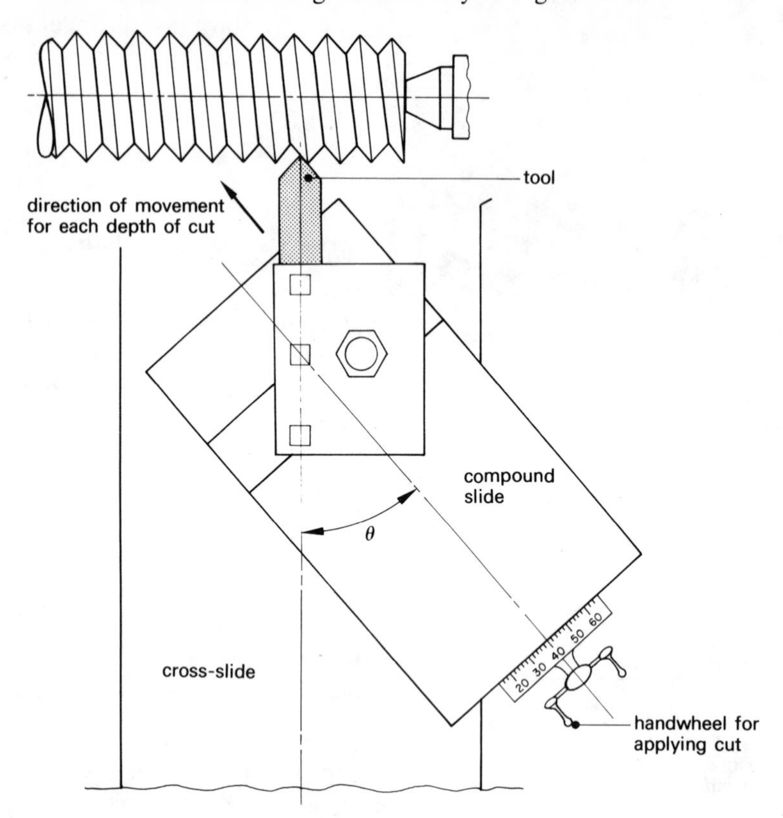

Figure 144 *Screw cutting using the compound slide*

The tool is set normal to the axis of rotation, as previously described. With the point just touching the work the cross-slide stop is set so that the cross-slide cannot be moved any further towards the workpiece. The speeds and feeds are selected as described previously and the tool is positioned ready to start cutting. The amount of cut is set using the compound slide. The leadscrew is engaged at the appropriate position and the first cut taken. The feed is continued by means of the cross-slide until the correct depth of cut is obtained. The major advantage of this process is that cutting is carried out on one side of the tool only. This produces a good surface finish. A top rake angle sloping away normal to the cutting face is advantageous to the final finish.

In both examples so far described the tool has been withdrawn from the work at the end of the cut. The cross-slide is returned to its initial position. Then the tool is fed into the new depth and the leadscrew engaged. An alternative method for returning the tool to its initial position is to reverse the direction of rotation of the spindle of the machine. Any backlash must be taken out of the gears before the tool is fed in to start the next cut. Another method on some machines is to reverse the direction of rotation of the leadscrew.

Self-assessment questions

18 Describe the primary function of a centre lathe.

19 Figure 145 contains a line diagram of a centre lathe. Complete the following table by naming the components numbered 1 to 7.

number	*component*
1	
2	
3	
4	
5	
6	
7	

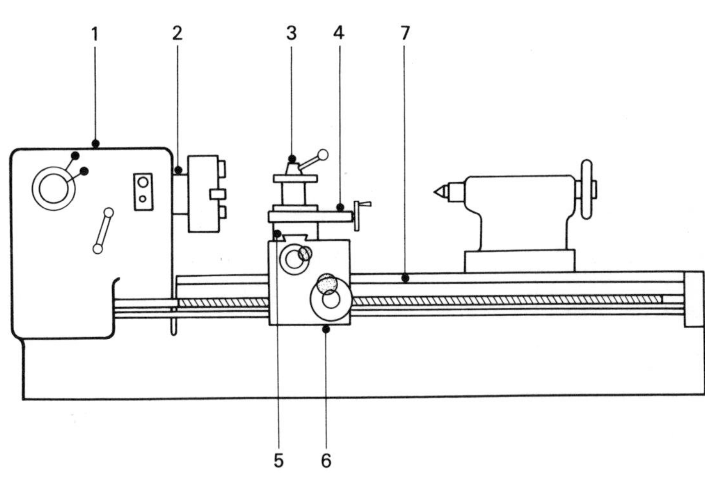

Figure 145 *Self-assessment question 19*

20 How is the speed of rotation of the workpiece regulated on a centre lathe?

21 Figure 146 shows a small piston used in a motorcycle engine. Sketch a form tool suitable for machining the grooves which accommodate the piston rings.

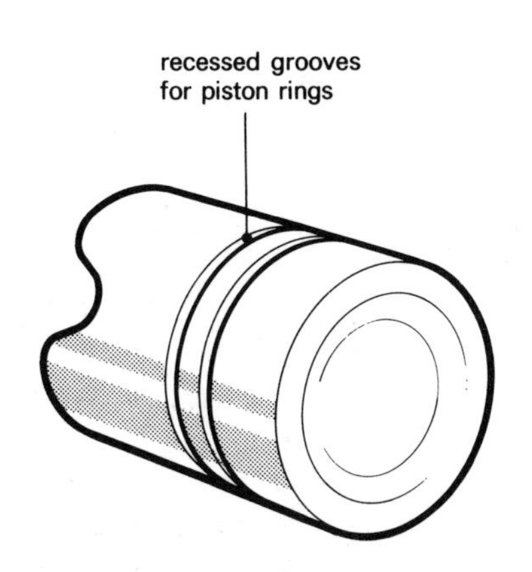

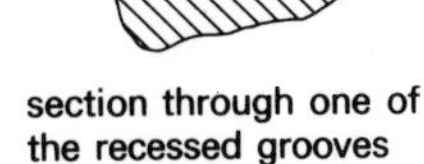

Figure 146 *Self-assessment question 21*

22 When would a four-way toolpost be used in preference to a single toolpost?

23 Sketch a workpiece held between centres. On the sketch show how the workpiece is rotated.

24 Describe two functions of the tailstock of a centre lathe.

25 In column 1 are listed four items which are to be machined on a centre lathe. In column 2 four work-holding devices are named.

	column 1	*column 2*
(i)	a hexagonal bar requiring a hole to be drilled in one end	(a) four-jaw chuck
(ii)	a large casting which has to be faced	(b) driving plate and centres
(iii)	a crankshaft which has to be turned	(c) face plate
(iv)	a long parallel shaft which is to be turned	(d) three-jaw chuck

Complete the table below by selecting the most appropriate work-holding device from column 2:

item	*most appropriate work-holding device*
(i)	
(ii)	
(iii)	
(iv)	

26 A taper may be produced using a compound slide. What type of taper is turned using this method?

Solutions to self-assessment questions

18 The primary function of a centre lathe is to produce cylindrical shapes by removing material.

19

number	*component*
1	headstock
2	spindle
3	toolpost
4	compound slide
5	cross-slide
6	saddle
7	bed

20 The speed of rotation of the workpiece depends on the speed of rotation of the spindle. The spindle speed is regulated by using a gear train that is situated inside the headstock.

21 Figure 147 illustrates a suitable form tool.

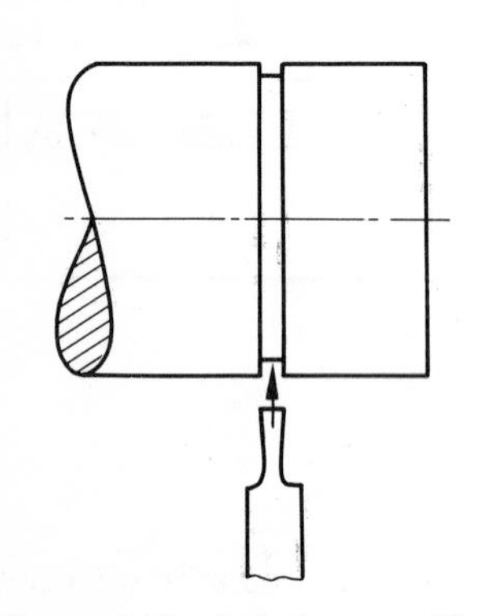

Figure 147 *Solution to self-assessment question 21*

27 In order to machine a taper using the tailstock offset, which item of the lathe provides movement for traversing the cutting tool?

28 What is the main difference between the two methods of screw cutting described in the text?

29 Give the two most common methods of ensuring that the tool locates the same groove for each cut during screw cutting.

30 What is the method used to set a tool to the correct position ready for screw cutting?

31 The movement of the screw-cutting tool along the lathe bed is important. How is the leadscrew, which drives the cross-slide, controlled?

Machining of simple components

Typical methods of machining simple components are given below. These are not the only procedures. Some of the components can be manufactured in a number of ways.

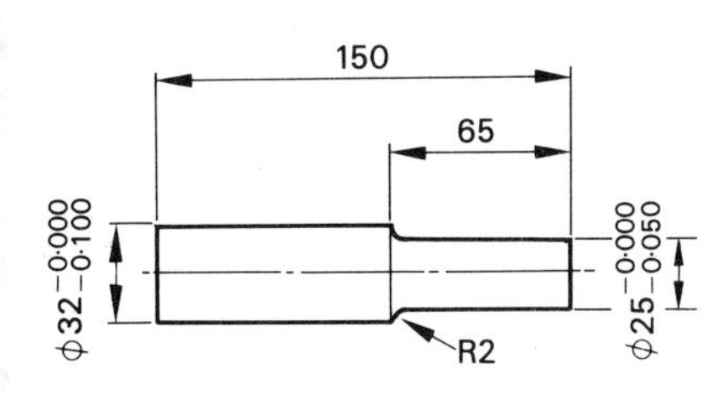

Figure 148 *Example 1: stepped pin*

Example 1

Stepped pin as shown in Figure 148.

A piece of mild steel with approximate dimensions ∅ 37 mm × 154 mm long is required.

A typical sequence is shown below.

1 A three-jaw chuck is placed on the lathe, a side tool is set in the tool-post and a drill chuck positioned in the tailstock barrel.

The three-jaw chuck is used initially to hold the round bar whilst the ends are prepared for machining between centres. The bar being turned would be of reasonable shape and size and suitable for holding in the three-jaw chuck.

A centre drill is small and needs to be fitted in a holding device. The drill chuck, which is suitable for holding the centre drill, has a morse taper which fits the barrel of the tailstock.

2 The bar of mild steel is held in the chuck with approximately 50 mm protruding.

3 The machine is started and the end of the bar is faced. The rotation of the chuck is stopped.

4 A centre drill is positioned in the drill chuck which has been fitted in the tailstock. The tailstock is moved near to the work and locked in position.

5 The chuck is rotated and by turning the handwheel at the end of the tailstock barrel, the end is centred. The rotation of the chuck is stopped, and the tailstock is removed from the work.

6 The work in the chuck is reversed and faced to length. The second end of the bar is centred.

The work now needs to be assembled between centres. This will support the workpiece, enabling reasonably large cuts to be taken, and also allow the appropriate length to be turned without distor-

Solutions to self-assessment questions

22 Consider turning a component that requires the use of more than one tool in its machining. If all the tools needed to complete the job were set in a four-way toolpost initially, the operation could be completed more efficiently than if a single toolpost were used. This becomes more important if more than one component is to be made. The overall setting time is greatly reduced by initially setting all the tools required in a multi-tool type of post.

23 See Figure 131, page 92.

24 A function of the tailstock is to position and locate tools and equipment that may be used for various operations carried out on the centre lathe. Another function is to support work whilst it is being machined.

25

item	*most appropriate work-holding device*
(i)	(d)
(ii)	(c)
(iii)	(a)
(iv)	(b)

26 As the compound slide is relatively short, the taper it can produce is also short. The length varies on different machines, but usually the maximum is about 100 mm. An advantage of this method is that a wide range of angles can be produced. This is particularly useful for steep tapers.

27 The saddle provides the movement, either operated manually or automatically via the feedshaft.

28 Using the first method the cut is put on normal to the axis of rotation. This means the tool cuts on both sides. A better method is to use the compound slide, set to an angle equal to half the included angle of the thread. By applying cut by means of the cross slide the tool is fed in at an angle, the tool thus cutting on one side only.

29 One method is to use the chasing dial. The tool is located at the appropriate line on the dial each time. The second method is not to disengage the leadscrew after a cut has been taken. If the leadscrew is reversed the tool progresses back to its starting position. Any backlash should be taken out of the gears.

30 A screw-cutting plate gauge is used. The tool is located in one of the grooves whilst the gauge is held against the work. This ensures that the tool is square to the axis of rotation.

31 The machines have auxiliary gearboxes beneath the headstock. By setting handles on the headstock an appropriate feed rate can be selected to give the necessary movement.

tion. The work may also be turned end to end during the process so that both diameters can be finished accurately to diameter and length.

7 The three-jaw chuck is removed and replaced by a driving plate and centre, on the spindle nose. A driving dog or carrier is fitted to one end of the bar. The drill chuck is removed from the tailstock, and replaced by a centre in the tailstock barrel.

8 A roughing tool is set in the toolpost, and the work is positioned between the centres of the machine.

9 The work is rotated and the material is turned to a diameter of approximately 33 mm, almost to the driving dog. The material is then finished turned to Ø 32 mm. Burrs are removed with a file.

10 The work is removed from the machine. Using packings to prevent marks a dog is fitted on the opposite end. The position of the length 65 mm is marked from the tailstock end, and the work re-assembled between centres.

11 Using another tool with a small ground nose radius, the tool is positioned so that the shoulder and radius can be produced without further setting.

12 The bar is turned down for a length of 64 mm approximately. When a diameter of approximately 25.25 mm has been reached, the travel is stopped. Any fine finishing of the shoulder may be done by manual control of the machine.

13 The tool is re-ground and set to cut the finished diameter. Regrinding is advisable because the sharpness of the tool helps to produce a better surface finish. When sufficient length has been turned to obtain a reading, the rotation is stopped and the diameter checked with a micrometer. If the size is within tolerance, the 65 mm length is finish turned. The traverse is stopped before meeting the shoulder and machining is finished by manual control of the machine. Any burrs on the end are removed with a file.

14 The finished component is removed from the machine and the dog is removed.

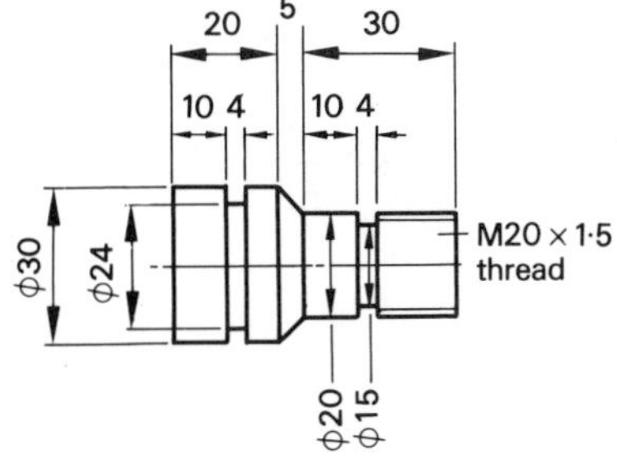

Figure 149 *Example 2*

Example 2

The component is shown in Figure 149.

A length of mild steel approximately Ø 32 mm is required. A long length would be advisable, since several components could then be produced from the bar.

A typical sequence of operations is as follows:

1 The material is set in a three-jaw chuck, leaving approximately 25 mm protruding from the chuck. The end of the bar is faced.

2 The material is extended to protrude approximately 65 mm from the chuck.

In both operations the amount protruding from the chuck is small with the result that no extra support is required. It is unlikely that excessive forces or vibration will be created.

3 The outside diameter is turned and finished to 30 mm.

4 Using a tool with a 45° approach angle the Ø 20 mm is turned and finished, producing the taper during the cutting operation.

5 A form tool 4 mm wide is positioned in the toolpost and the two grooves are machined.

6 Using a special form tool mounted in the toolpost on the compound slide, set to half the included angle of the thread, the screw is cut as described in the text (pages 97–99).

7 The component is parted off 20 mm from the taper.

8 The bar may be extended from the chuck and operations 1 to 7 repeated.

Example 3

The component is shown in Figure 150.

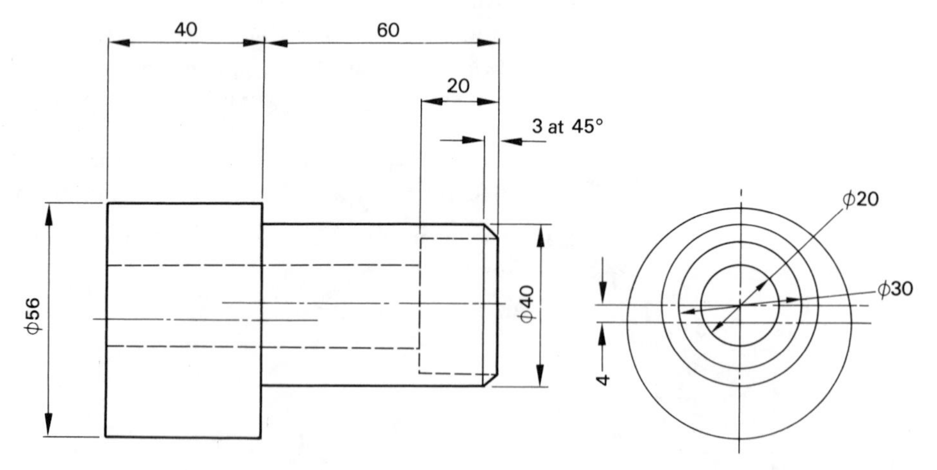

Figure 150 *Example 3*

A piece of mild steel approximately Ø 60 mm × 110 mm long is required.

Since part of the component is offset, machining must be carried out in a four-jaw chuck. To prevent changing the work holding during the operation sequence, the four-jaw chuck is to be used throughout. It will also be suitable if either hot-rolled or cold-rolled material is used.

1 The bar is set in a four-jaw chuck using a dial test indicator. Approximately 60 mm is allowed to protrude from the chuck.

2 The end of the bar is faced.

3 The Ø 56 mm is rough turned and finished turned for a length of approximately 50 mm.

4 The component is removed from the chuck and the second end is faced.

5 The component is again removed from the chuck and the offset diameter is marked on the face just machined.

6 The component is reset in the four-jaw chuck offsetting the component so that the diameter marked in **5** runs true.

7 The bar is faced to length and a centre mark produced on the end.

8 The Ø 40 mm is rough turned and finish turned, completing the final part by manual control of the machine to produce the square shoulder.

9 A drill chuck containing a Ø 6 mm drill is placed in the tailstock. A drill chuck is required to hold the drill as in Example 1.

10 A pilot hole is drilled through the whole component.

11 The drill chuck is removed and a drill is positioned in the barrel of the tailstock (a suitable size would be Ø 20 mm.)

12 A Ø 20 mm hole is drilled through the component.

13 A boring tool is set up in the toolpost and the toolpost is positioned to enable the Ø 30 mm hole to be bored.

14 The hole Ø 30 mm is bored, taking care at the end to ensure the square corner.

15 Using a form tool the chamfer on the end is produced.

16 Any burrs are removed, the machine is stopped and the component is removed.

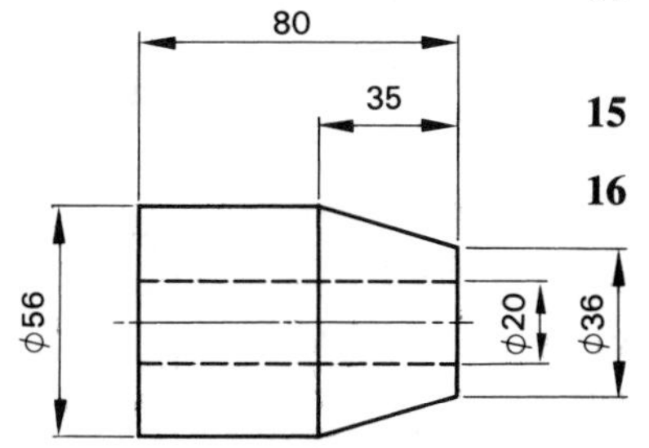

Figure 151 *Self-assessment question 32*

Self-assessment question

32 Describe a suitable sequence of operations for machining the component shown in Figure 151.

After reading the following material the reader shall:

8 Describe the function and use of the shaping machine.

8.1 Identify the component parts of a typical shaping machine (e.g. ram, body, table, drive system).

Until the beginning of the nineteenth century engineers avoided, wherever possible, incorporating flat surfaces in their designs. At that time these surfaces were very difficult to produce. If flat surfaces were essential, then they were produced by chipping (using a chisel), by filing or by scraping. A machine tool, the 'jigger', was introduced in 1836 by James Nasmyth and was the first in a series of steps that led to the shaping machine of today. In operation the tool

Solutions to self-assessment questions

32 A typical sequence of operations is outlined below:

A piece of mild steel approximately Ø 60 mm × 86 mm long is required.

1 The material is placed in a three-jaw chuck or a four-jaw chuck as described in the previous examples.

A chuck is required to hold the component because of the hole which is to be machined through the centre. Using centres would involve several changes of work-holding equipment. The three-jaw chuck may be used for cold-rolled bar and the four-jaw chuck for hot-rolled bar.

2 The end is faced and centre drilled.

3 Using the chuck and the tailstock centre for support, the bar is turned down for a length of 45 mm to a diameter of 56.5 mm. It is then finish turned to Ø 56 mm.

The tailstock is used to support the material to prevent any possible distortion during the machining operations. In this application, machining may be carried out successfully, if care is taken, without the use of a support. However, the use of the centre will ensure better holding and better accuracy.

4 The component in the chuck is reversed using packings to prevent markings. Approximately 25 mm is held in the chuck.

5 The bar is faced to length and centre drilled.

6 The compound slide is set to an angle of 16° to the axis of rotation.

7 The taper is turned using the handwheel on the compound slide to move the tool across the surface.

8 A drill Ø 8 mm is positioned in the barrel of the tailstock.

9 A pilot hole is drilled through the centre of the component.

10 The Ø 8 mm drill is removed and replaced with a Ø 20 mm drill.

11 A Ø 20 mm hole is drilled through the component.

12 Any burrs are removed, the machine is stopped and the finished component is removed.

in the new machine appeared to dance to and fro; this is why the name 'jigger' was given to it. This motion is retained in the modern shaping machine.

The main function of a shaper is to generate flat surfaces by combining a linear movement of the cutting tool with a linear movement of the workpiece. These two movements are at 90° to one another. (See Figure 153.)

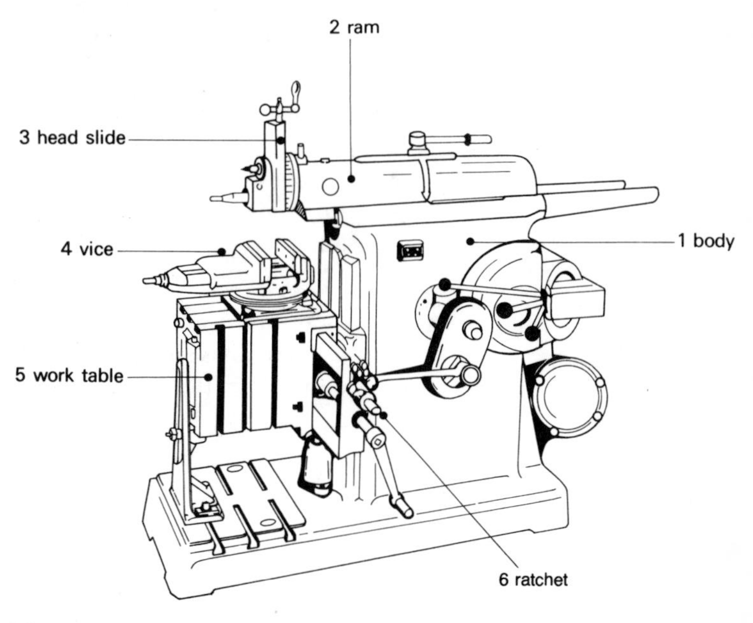

Figure 152 *The shaping machine*

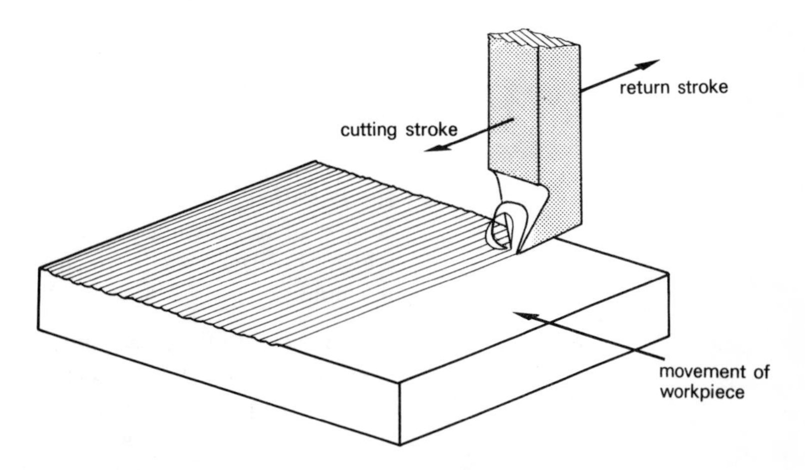

Figure 153 *Shaper tool and work movement when generating a flat surface on a shaping machine*

Using this cutting process means that only on the forward stroke of the tool is material removed; the return stroke is non-productive. Because of this, for an equal power consumption, other machine tools (e.g. milling machines) remove a greater volume of material. The shaping machine has the advantage, however, of being easily set up for 'one-off' jobs that occur, for example, in tool rooms. The shaping machine is a versatile piece of equipment used for many purposes. Because of its relatively simple design, a shaper is a relatively cheap machine.

The shaper basically does five things:

1. It holds the work – in a vice on the table.
2. It holds the tool – in the head slide.
3. It moves the tool backwards and forwards – using the ram.
4. It provides a vertical feeding movement to the tool – in the head slide.
5. It provides a transverse movement to the workpiece.

There are a number of other machine tools which generate flat surfaces. These include the milling machine, the surface grinder, the centre lathe, the drilling machine (when spot-facing), and the planing machine.

The block diagram (Figure 154) shows the main parts of a shaping machine.

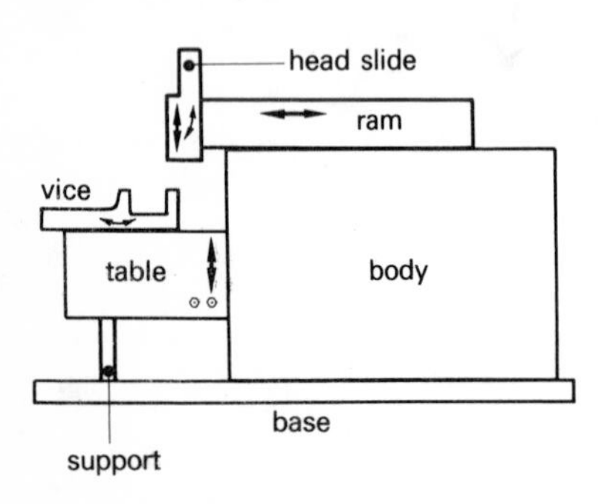

Figure 154 *Block diagram of a shaping machine*

The body

During the operation of a shaping machine, large forces come into play. For this reason it is important that the construction of a shaper is not only strong, but is also rigid, thus avoiding distortion of the machine parts.

The *body* (item 1 in Figure 152) is made from cast iron and, as with the lathe bed, takes the form of a box-type section. The hollow section of the casting contains the gearbox and drive mechanism which are driven by an electric motor situated at the rear of the machine.

Machined on the body casting are two sets of bearing surfaces in the shape of *dovetail* slideways (Figure 155). It is important that these bearing surfaces are machined at 90° to one another as this ensures the accuracy of any workpiece produced. The horizontal bearing surfaces on top of the body casting are for the location of the ram (item 2 in Figure 152). The vertical bearing surfaces on the front face of the body locate the worktable.

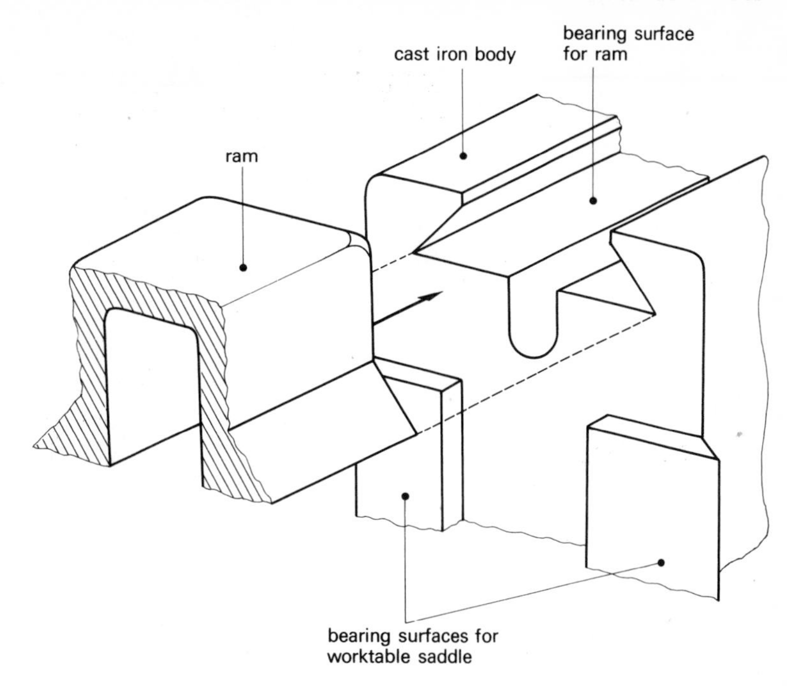

Figure 155 *Shaping-machine guideways*

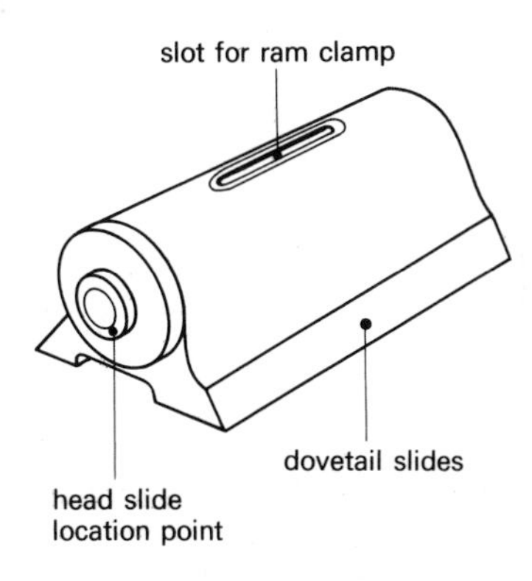

Figure 156 *The shaping-machine ram*

The ram

The main function of the ram is to provide accurate movement of the cutting tool. Like the body, the ram is made of cast iron. This provides the rigidity essential. The ram is located by the horizontal dovetail slideways situated on the top of the body casting. These slideways mate with the dovetails on the ram casting (Figure 156). The motion of the ram is a reciprocating (backwards and forwards) movement, the ram being activated by the driving mechanism located in the body.

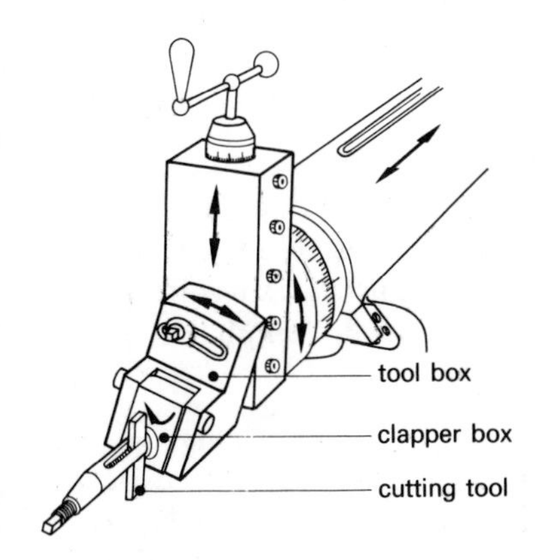

Figure 157 *The head slide*

The head slide

The head slide assembly (item 3 in Figure 152) is located on the front face of the ram casting. Figure 157 shows a head slide which can rotate on its mounting. It can also move vertically, enabling the depth of cut to be altered. The forward stroke of the tool is used for the cutting operation. The backward stroke returns the cutting tool to its original position, and the cycle starts again. During the return stroke the tool is allowed to run free due to the small angular movement of the clapper box which allows the tool to lift and therefore slide back over the workpiece (Figure 158).

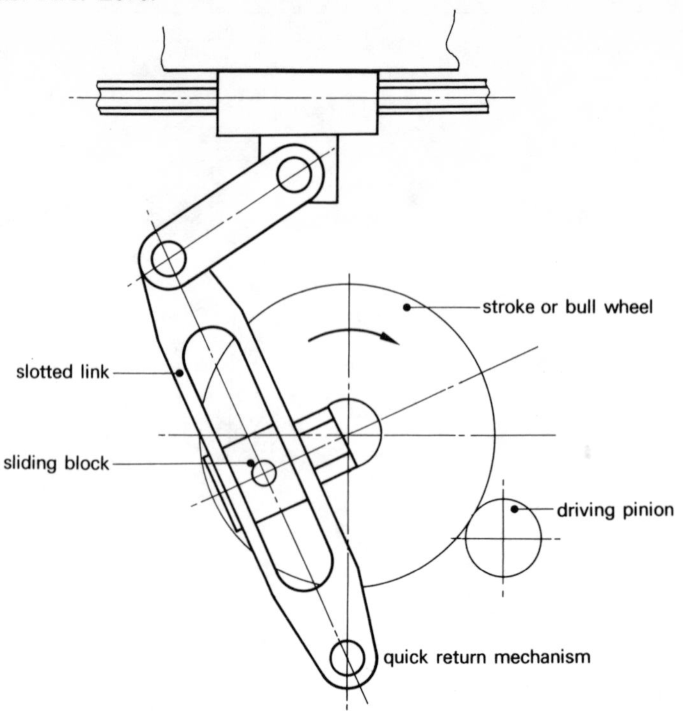

Figure 158 *Quick-return mechanism*

The table

A workpiece to be machined on a shaping machine is held in a vice situated on top of the table (Figure 152). The table is usually a strong, box-like casting. The vertical slideways on the body of the shaper permit the raising or lowering of the table, whilst maintaining the necessary geometrical accuracy.

After the table has been positioned vertically, it is clamped to prevent deflections whilst cutting is taking place. On heavy-duty shapers a support is provided (Figure 154). This gives rigidity, and prevents deflection of the workpiece when the cut is heavy.

Figure 153 shows the transverse movement of the workpiece necessary to produce a plane surface. The movement of the table where the workpiece is held is normally brought about by an automatic or mechanized action. It is also possible to cause a transverse movement of the table by hand feeding; this, however, is both tedious and laborious.

The surface produced by a shaping machine is not of a high quality. In order to ensure that the best possible surface is produced, before the finishing cut is taken the tool should be removed and sharpened. A light finishing cut (with a good supply of cutting fluid) gives the workpiece a flat finish.

The drive system

A 'quick-return' mechanism causes the reciprocating movement of the ram.

One function of this mechanism, as its name suggests, is to return the ram to the start of the cutting stroke faster than it moves the ram during the cutting stroke. This mechanism keeps the non-cutting time to a minimum.

In a shaper the drive is applied to the tool, the work being traversed slowly beneath the tool. The ram carrying the cutting tool is actuated by a slotted-link quick-return mechanism. Power from an electric motor drives the stroke (or bull) wheel. This in turn actuates the slotted link by means of a sliding block. The reciprocating motion of the ram means that the cutting tool is accelerating at the beginning of the cutting stroke and retarding at the end of this stroke. In a number of cases this results in a significant variation in the cutting speed during the stroke.

Safety

Safety is important when using any machine tool; the shaper is no exception. The workpiece must *never* be touched whilst the machine is operating, and the operator must not stand in front of the ram. When carrying out adjustments to the workpiece, measuring or changing settings, the operator must switch off the electrical supply to the machine. Metal chips from a shaping machine tend to fly in all directions, so it is important to wear goggles or safety glasses.

Whenever possible, the vice on a shaping machine should be fitted as shown Figure 152: that is, the vice should be positioned so that the jaws are at an angle of 90° to the direction of cutting. If the jaws of the vice are parallel to the direction of cutting, then only friction forces (between the workpiece and the jaws) are holding the workpiece in position. This may be dangerous.

Self-assessment questions

33 Draw a sketch to show the relative movements of
(i) the cutting tool and
(ii) the workpiece on a table
when producing a flat surface using a shaping machine.

34 Study diagrams (a) and (b) in Figure 159, and then state which is the cutting stroke and which is the return stroke.

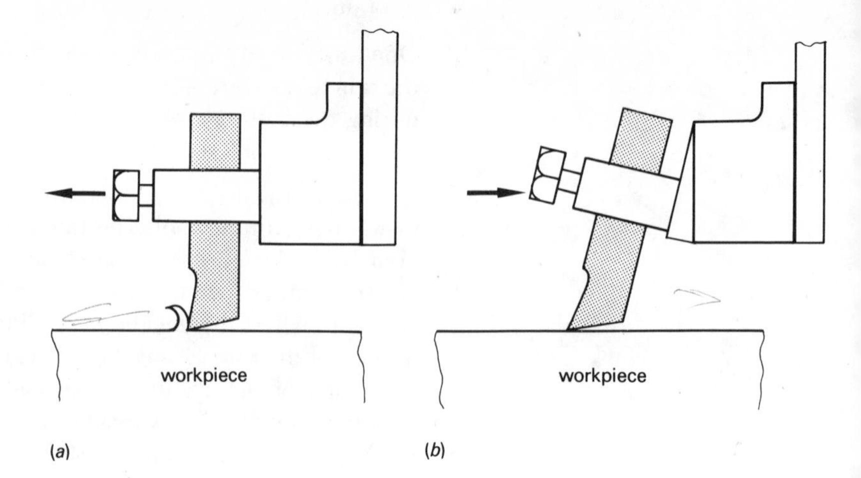

Figure 159 *Self-assessment question 34*

35 What are the main functions of the body of a shaping machine?

36 What name is given to the component that provides accurate movement of the cutting tool?

37 What assembly is used to adjust the depth of cut?

38 What is the purpose of the clapper box?

39 Figure 158 (page 110) shows the ram in position at the beginning of the cutting stroke. Make a line sketch showing the position of the slotted link when the ram has just completed the cutting stroke.

After reading the following material, the reader shall:

9 Describe the selection and maintenance of cutting tools and the need for cutting fluids.

9.1 Identify the relevant angles on common cutting tools.

9.2 Select appropriate tools for various machining operations such as producing holes of various standards.

9.3 List the advantages of cutting fluids.

For a machine tool to cut material efficiently, it is important that the correct cutting tools in a suitable condition are used. The type of tool, size of angles and shape of the tool can vary quite appreciably. The correct combination of these factors together with speeds, feeds and setting, contribute to the efficient production of components.

It is only during the last sixty years that the metal-cutting process has been understood. This, together with new materials, has enabled cutting tools to be manufactured capable of withstanding the demand for higher production.

A cutting tool showing all the relevant angles is detailed in Figure 160.

The diagram shows the two main types of angle. These are known as:

(i) *rake angles*, and
(ii) *clearance angles*

The rake angles are important to the actual cutting process. They determine the type of chip which is produced and, as a result, the quality of the finish on the component. Clearance angles, however,

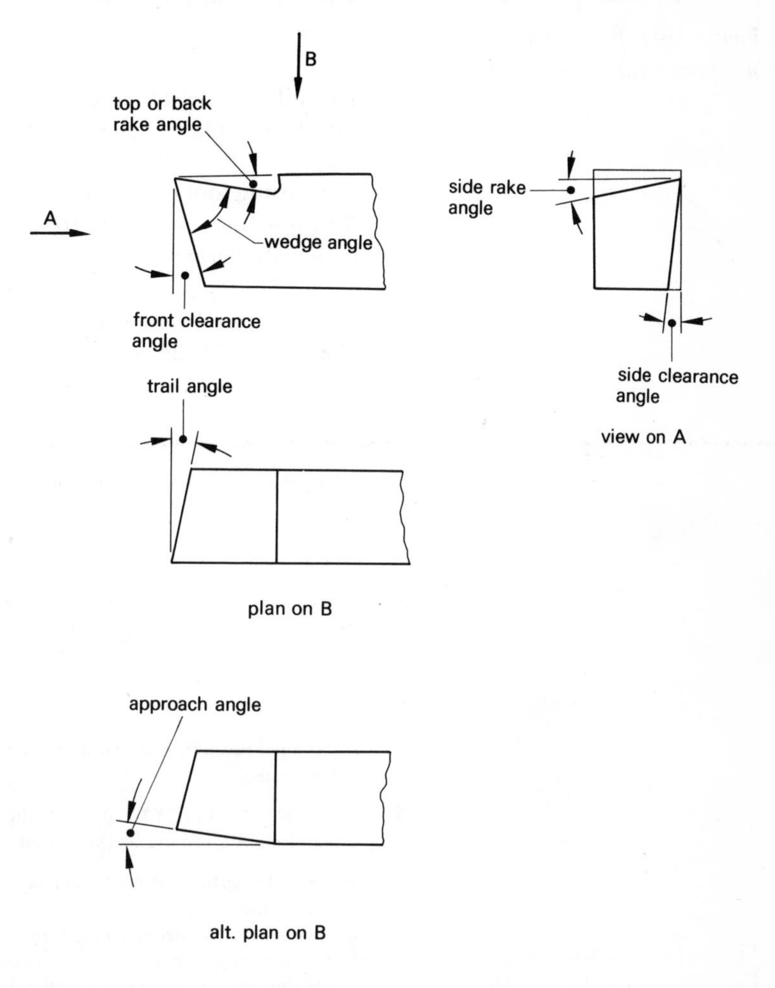

Figure 160 *Cutting-tool angles*

material being cut	rake angle
cast iron	0°–14°
brass	0°–14°
tool steel	8°–10°
mild steel soft steel	20°–27°
copper	30°–35°
aluminium	35°–40°
perspex	40°–50°

Figure 161 *Rake angle for machining various materials*

are provided to prevent the tool rubbing the workpiece. This rubbing generates unwanted heat, causing inferior work finish, and in some cases, damage to the tool.

Typical top rake angles for cutting different materials are listed in Figure 161. These are now regarded as the most efficient, taking into account such items as forces, power required, cutting speeds, cutting-tool life, surface finish and properties of the material being cut. The angles have been arrived at from observations made during actual cutting operations. The theory of cutting tools has been advanced by studies carried out both in research and production laboratories.

Incorrect positioning of the cutting tool increases or decreases the rake angle from its correct value with a resultant loss in efficiency. The effect on rake angle of an incorrect tool height is shown in Figure 163.

The differences obtained in cutting mild steel with tools having various top rake angles are shown in Figure 164.

The cutting tool may take different forms. The most common are:

1 A tool made completely from high-carbon steel. The appropriate angles are formed by grinding.
2 A tool-holder containing a clamped cutting tool. The cutting tool material may be high-carbon steel, tungsten carbide or ceramic. The type of holder can vary.

Figure 162 *Solution to self-assessment question 39*

Solutions to self-assessment questions

33 Figure 153 (page 107) illustrates the appropriate movements.

34 (a) is the cutting stroke and (b) is the return stroke.

35 It provides a rigid structure in which are the location surfaces (dovetail slideways) for the ram and worktable. It also houses the drive mechanism and motor.

36 The ram.

37 The head slide.

38 The clapper box allows the cutting tool tip to ride over the workpiece surface on the return stroke.

39 See Figure 162. Line *XY* represents the slotted link at the beginning of the cutting stroke. Line *XZ* represents the slotted link at the completion of the cutting stroke.

Note that the cutting stroke takes place while the stroke wheel completes arc *a*. The return (non-cutting) stroke takes place in the period that the stroke wheel turns through arc *b*. The stroke wheel turns at a steady angular speed. Since arc *b* is less than arc *a* it follows that the return stroke takes less time than the cutting stroke. This is why this type of linkage is called a quick-return mechanism.

correct position and top rake angle

tool set at correct height

new top rake

this positioning reduces the top rake. In the exaggerated position shown it has become a negative angle

A

this positioning increases top rake; there is also a tendency for the front of the tool to rub at area A

Figure 163 *Effect on top rake angle of tool height*

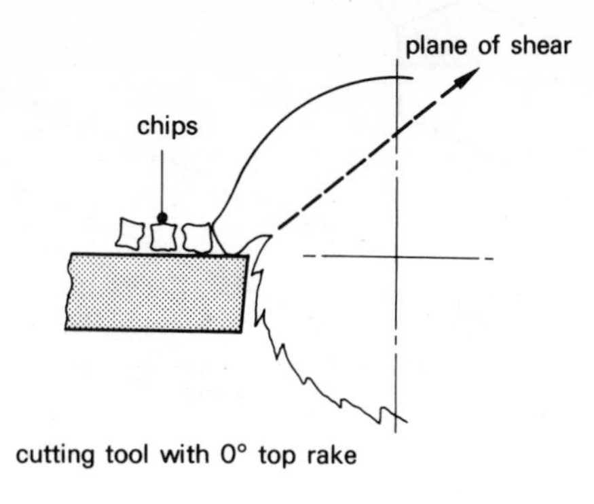

RESULTS:
very poor finish,
tool point is not in contact with work
material tears
long shear plane

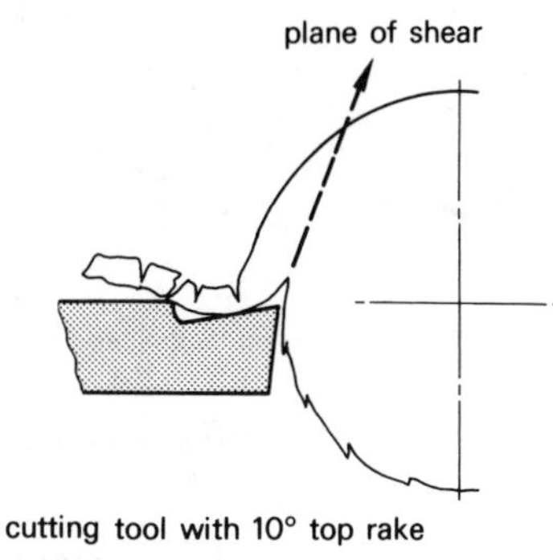

RESULTS:
an improved finish but still poor quality,
tool point is in contact part of the time
material shears instead of tearing
a shorter shear plane

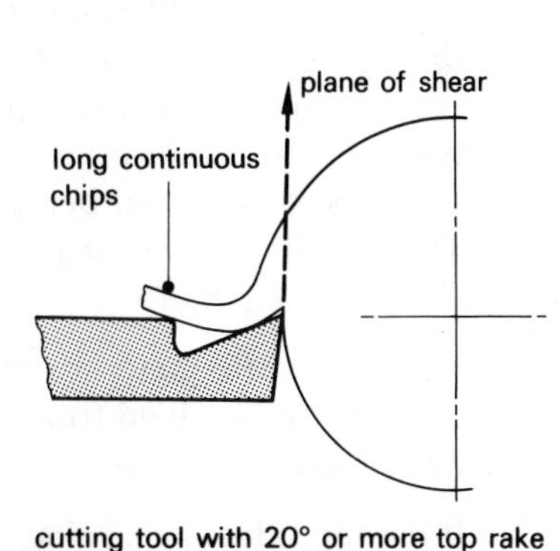

RESULTS:
good finish
material now flows
small shear plane, therefore less energy is required–less distortion

Figure 164 *Effect of rake angle on cutting of mild steel*

3 A tool-holder having a tool tip brazed in position. Once again there is a variation in tool materials.

Some tools and tool-holders are shown in Figure 165.

When a tool is cutting material large forces are set up. As the metal being removed is compressed, a zone of severe friction is set up near the end of the tool. The tool must withstand both the force exerted and the high friction force. The friction generates heat which may

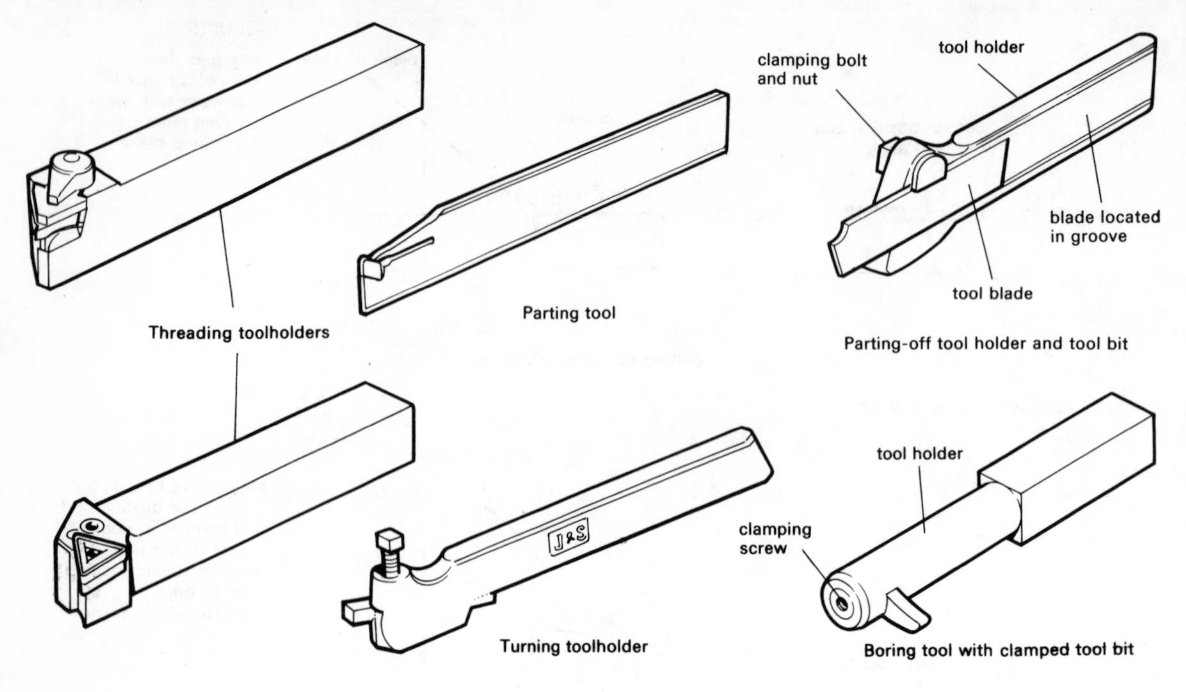

Figure 165 *Tools and tool-holders*

soften the tool. This may result in a blunt edge which rubs instead of cutting. The compressive force on the material being cut increases to a point where shear begins to take place within the material. The shear plane which is set up is almost normal to the tool face. The type of chip produced is determined by

(i) The type of material being cut
(ii) The conditions in which the cutting tool is operating
(iii) The top rake angle

Usually, brittle materials produce small chips, whilst a continuous chip can be obtained from a ductile material. The basic conditions during cutting are shown in Figure 166.

If the top rake angle is increased, both the forces on the tool point and the power required are reduced. However, an increase in the rake angle above the values quoted in Figure 161 shortens tool life. This is because the wedge angle of the tool becomes progressively smaller and the tool material is unable to withstand the forces and heat generated during cutting. Some cutting tool shapes are shown in Figure 167.

The single-point tool tends to produce a finish inferior to a round-nosed tool. However, if the radius becomes too large on the round-nose tool, chatter may occur so the finish is not as good. This is due to the large length of tool in contact with the material during cutting.

Figure 166 *Conditions existing during cutting. The diagram shows a tool on a shaping machine; however, the principle applies to all cutting tools*

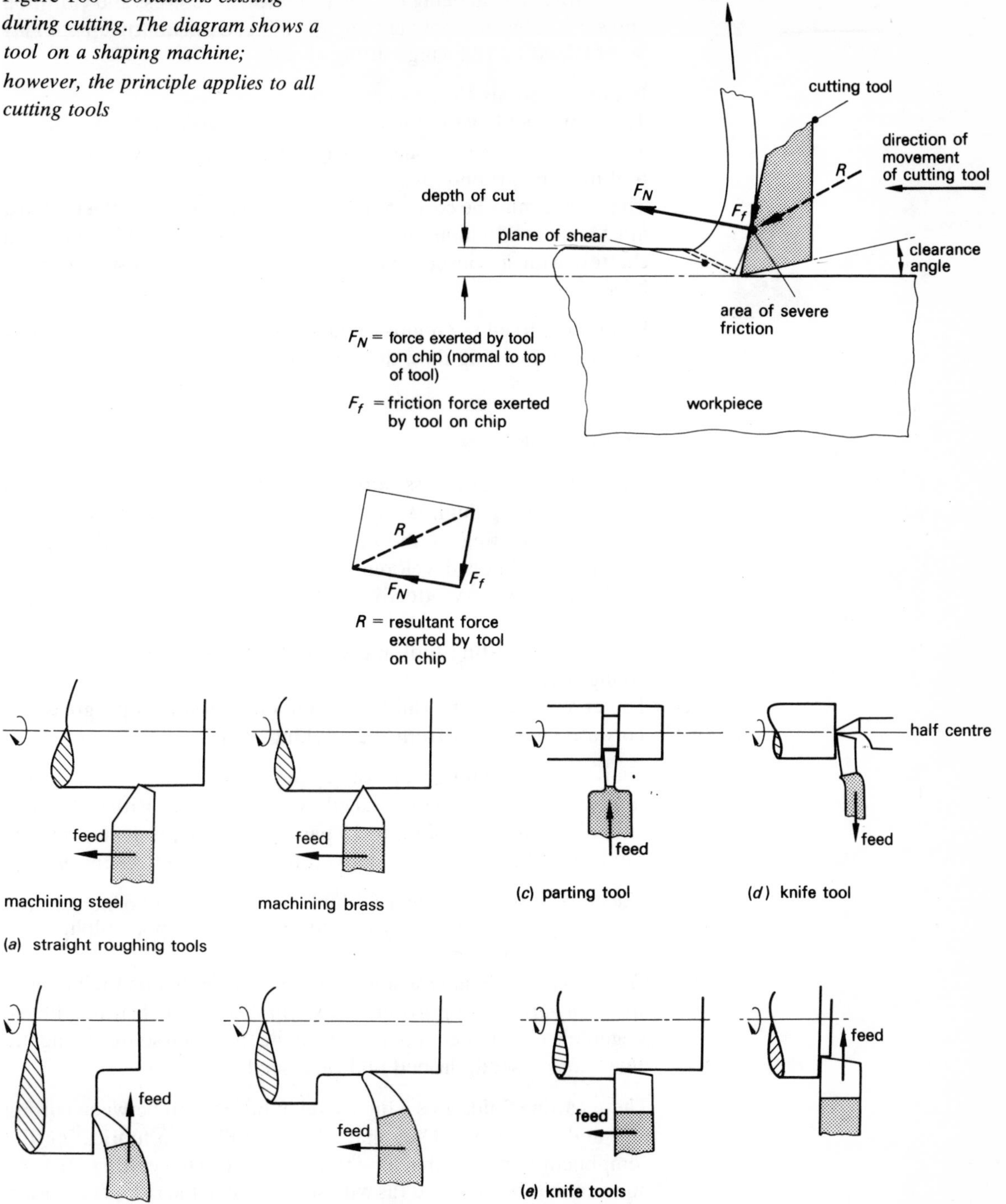

Figure 167 *Profiles of single-point cutting tools*

The extra material being removed creates more stress and vibration and sometimes the tool may rub and not cut. An improvement may be obtained by reducing cutting speed.

Important points to remember are:

1 The tool must have the correct degree of hardness.
2 The tool must be kept sharp. Any defects which may occur on the tool must be ground away.
3 The setting must be correct. The overhang of tool from the toolpost must be kept to a minimum. Large overhangs lead to vibration and chatter, with a consequent reduction in the standard of surface finish.

In addition to all the points discussed it is also advisable to employ the use of a cutting fluid during cutting processes.

Cutting fluids

To aid the cutting process, cutting fluids are used. Initially water was used as a cutting fluid. It kept the cutting tool and workpiece cool, but caused rusting of the cutting tool, workpiece and machine tool parts. Major developments have taken place and cutting fluids are now very effective. Basically a cutting fluid has three functions:

(i) To keep the cutting tool and workpiece cool whilst machining is taking place
(ii) To act as a lubricant whilst a cutting operation is in progress
(iii) To wash chips away from the working area

There are two main groups of cutting fluids:

(a) *Soluble oils*. These are mineral oils which are mixed with water (in a ratio of about 1 part of oil to 25 parts of water), producing a white emulsion. Mineral oils are derived from crude petroleum oils.

The lubricating properties of soluble oils are limited; to increase the lubricating action a variety of additives, based upon sulphur and chlorine, can be used.

(b) *Cutting oils*. Both mineral and vegetable oils (either on their own or mixed in various proportions) make up this group of cutting fluids. Vegetable oils are derived from plant life, the main source being the olive, cotton seed, linseed and rape seed.

These cutting fluids give a high level of lubrication between cutting tool and workpiece. They may also be utilized in lubricating the component parts of a machine tool adjacent to the cutting area. As no water has been added (as with soluble oils), the risk of corrosion occurring through their use is minimal. Because of this they are extensively used on automatic machine tools and expensive gear cutting machines.

Self-assessment questions

40 When turning a ductile material the chip produced tends to be continuous.

TRUE/FALSE

41 Given the cutting tools illustrated in Figure 168, which one is used for machining:
(a) Mild steel
(b) Brass
(c) Aluminium
(d) Tool steel

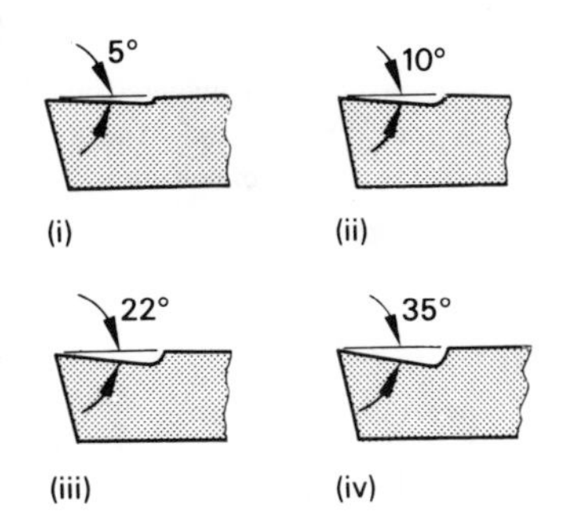

Figure 168 *Self-assessment question 41*

42 On Figure 169 indicate top rake, front clearance angle and the wedge angle.

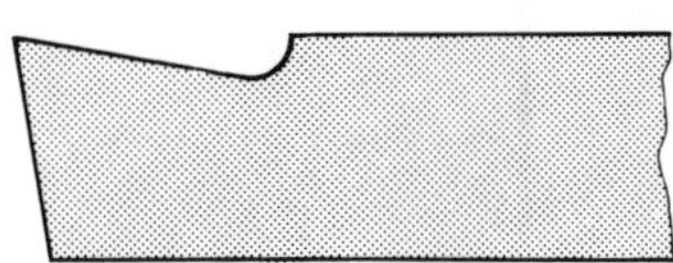

Figure 169 *Self-assessment question 42*

43 What is the effect of using a cutting tool which does not have a front clearance angle?

44 In column 1 are listed four machining operations to be carried out on a centre lathe. In column 2 three types of cutting tool are named.

column 1	*column 2*
(i) to separate a machined component from the stock material	(a) straight roughing tool
(ii) to turn a straight cylinder	(b) side tool
(iii) to surface up to a radiused corner or shoulder	(c) parting tool
(iv) to produce a groove in the surface of a component	

Complete the table below by selecting the most appropriate cutting tool

machining operation	*tool type*
(i)	
(ii)	
(iii)	
(iv)	

45 List three reasons for using a cutting fluid.

46 Name the two main groups of cutting fluids used in engineering workshops

Solutions to self-assessment questions

40 True.

41 (a) (iii)
(b) (ii) or (i)
(c) (iv)
(d) (ii)

42 See Figure 160, page 113.

43 A cutting tool without front clearance rubs the work being machined. Unwanted heat is generated; the surface finish of the machined component is affected. Another effect of the cutting tool rubbing the workpiece is to increase the power consumption during the cutting operation.

44

machining operation	*tool type*
(i)	(c)
(ii)	(a)
(iii)	(b)
(iv)	(c)

45 The reasons for using a cutting fluid include:
(a) To cool cutting tool and workpiece
(b) To act as a lubricant
(c) To wash away chips formed during the machining operation

46 (a) soluble oils
(b) cutting oils

Topic area: Fastening and joining

After reading the following material, the reader shall:

10.1 Distinguish between soldering and brazing.
10.2 Describe the principles involved in soft soldering, including electrical applications.
10.3 Describe the principles involved in silver soldering and brazing.

Soldering and brazing are two similar processes which are used for joining materials. *Soldering* utilizes a much softer material than brazing and creates a weaker joint, which is suitable only if (i) there are no vibrations on the joint and (ii) the operating temperatures are low. *Brazing* in contrast provides a tough and strong joint.

Soft soldering is a quick method of joining two items where strength is not important. It is particularly suitable for joints in electrical conductors.

The basic principle of soft soldering is that the molten solder being used to join two metals should 'wet' the surfaces of the two metals. This means that some atoms in the solder must combine with some atoms on and near the surface of each piece being joined, thus forming an alloy (Figure 170).

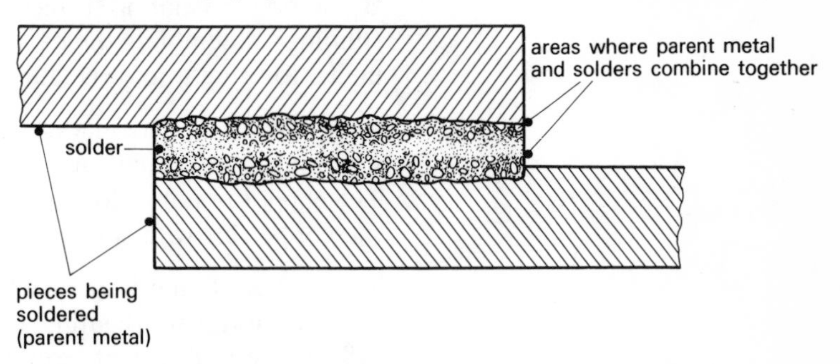

Figure 170 *Soft soldering*

Since soft solder has a low melting temperature (between 180 and 220 °C), there is no danger of melting the two metals which are being joined; also there is little or no effect on the structures of these materials.

To ensure that the solder 'wets' the surfaces, it is essential that the

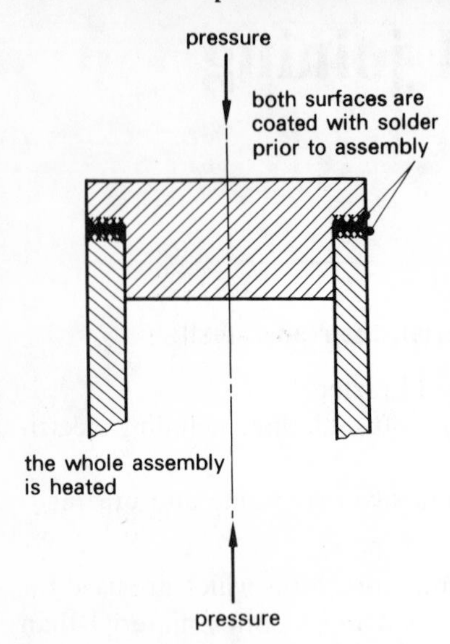

Figure 171 *Soldering*

surfaces are perfectly clean, i.e. free from dirt, grease or oxides. If any contamination is present, the molten solder forms globules on the surface instead of 'wetting', with the result that a bond does not occur between the two pieces. Although the metals may be cleaned and polished, immediately heat is applied an oxide film forms and the solder cannot create the joint.

To overcome this problem and assist in maintaining the necessary clean joint, a flux is used. There are two basic types of flux: one purely protects after a surface has been cleaned (non-active); the other protects and also cleans (active). The most widely used flux is zinc chloride (called 'killed' spirits). Another common flux is hydrochloric acid. Both of these fluxes are of the active type. The chief problem with them is that they leave behind a residue which may cause corrosion. When the joint cannot be washed after soldering, it is advisable to use one of the non-active (organic) type of fluxes such as tallow, resin, vaseline, etc.

For soldering electrical joints where the work is usually light and continuous, it is usual to employ a solder within which a flux is incorporated.

Production of soldered joints

Joints which are soldered may be produced in a number of ways:

(a) The parts to be joined are heated, coated with flux and the solder is added.
(b) Assembled and fluxed parts are dipped into molten solder
(c) By using a soldering iron
(d) By coating the mating surfaces of the parts to be joined with solder (tinning), assembling the parts and heating them so that the solder melts, forming a joint (Figure 171)

Most small joints are made by using a soldering iron. The bit of the soldering iron is usually made of copper (a good conductor of heat). The copper readily alloys with tin and this facilitates the coating of the copper bit with solder.

Types of soldering iron vary from solid types (of varying shapes) which require flame heating, to the modern types which are electrically heated. This text mainly concentrates on the solid, flame-heating type.

The materials to be joined must be prepared so that they are clean and free from dirt, grease or oxide films. This is usually achieved by either filing or rubbing with dry emery cloth the surfaces to be joined. After cleaning, a coating of flux is immediately applied. The joint is now ready for the application of the solder. During the preparation of the joint it is usual to heat the soldering iron bit to a

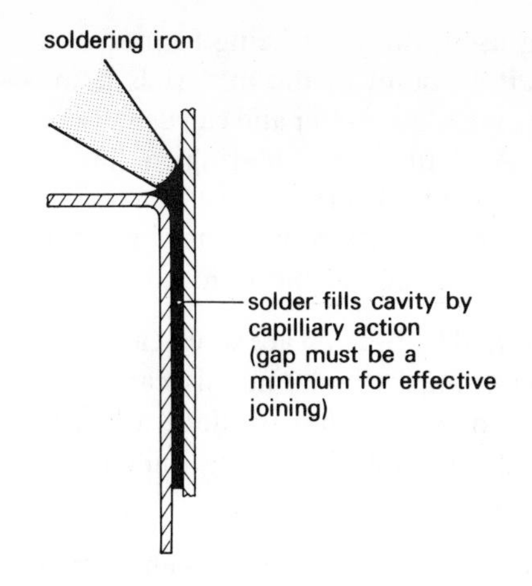

Figure 172 *Capillary action in soldering*

temperature which causes the solder to melt easily. When it is hot enough the iron is removed from the heat and filed.

The iron, after being filed, is coated with flux to maintain the cleanliness. It is then dipped in solder (tinning). A thin film of solder adheres to the iron provided that it has been correctly cleaned. The iron bit is then applied to the joint and drawn along at a slow, even rate. Heat from the iron melts the solder and heats the workpiece to a temperature which allows the molten solder to flow into the joint as it cools. Extra solder may be added as the iron is being drawn along to ensure a continuous joint. The heat from the iron makes the flux give off a gaseous shield. This prevents atmospheric attack during the soldering process. A good joint has a small amount of solder with excellent adhesion.

Electrically heated soldering irons are most suitable and widely used for light-duty applications such as electrical connections, etc. Since a small soldering iron is required for this type of application the flame-heated type lose heat very rapidly after withdrawal from the heat, making the production of suitable and satisfactory joints very difficult. For large-quantity production of soldered joints the electrically-heated types have advantages.

Many joints depend on the capillary action of the solder to create a good joint. Capillary action means that liquids spontaneously elevate or depress when they come into contact with very thin gaps or openings. A soldered joint is produced so that the parts fit close together leaving a very small gap. When molten solder is applied the action causes it to fill the gap, forming a joint. The principle of this is shown in Figure 172.

The usual clearance between surfaces which have not been tinned is approximately 0.12 mm. This may be reduced quite considerably if the surfaces have been previously tinned.

It is better to have the soldering iron too hot rather than too cold provided that it is not red hot. Similarly, it is advisable to warm the pieces being joined rather than operate with cold materials.

Typical soft solders are usually alloys of lead and tin. The melting range of these alloys is between 182 °C and 250 °C.

Brazing

Brazing is a process based on a similar principle to soft soldering. Brazing uses an alloy of copper and zinc (the minimum copper content is 50%) as the filler rod and is used for joining metals including dissimilar metals where a joint is required which is stronger, harder and more rigid than joints produced by soft soldering. The operating temperatures range between 600 °C and 900 °C depending upon the alloy being used, the joint being formed at a temperature lower than the melting point of the materials being joined. The alloy used for brazing is called spelter and can usually be obtained in strip or granules. As with soft soldering, a flux is required; borax is used most extensively. Borax dissolves the oxides and, under the effect of a red heat, vitrifies (becomes glass-like) and forms a protective layer above the surface of the joint.

The materials most commonly joined by brazing are steel, cast iron, brass, bronze and copper. When joining dissimilar metals, care must be taken to choose a filler rod which does not produce a brittle compound when it combines with the materials being joined.

Once again the important point is to ensure the joint is clean. Flux is then added and the joint heated, usually with an oxy-acetylene blowpipe. The temperature of the materials being joined must be high enough to melt the spelter. When a sufficiently high temperature has been reached spelter, together with more borax, is added to the joint. As the spelter melts it is drawn into the gap by capillary action. The end-product is a strong but ductile joint. Since capillary action is necessary, the clearance between the parts should be carefully controlled if maximum strength is required. After forming the joint, the materials must be allowed to cool slowly since any quenching may create cracks and distort the material.

If a flux is used which employs a chemical action, the materials must be thoroughly washed after brazing to prevent corrosion.

In addition to hand brazing with an oxy-acetylene blowpipe, the process may be semi-automatic for large quantities; for large components, heating may be carried out inside a furnace. When this method is used the parts are assembled including flux and spelter, clamped together and placed in the furnace. Several components may be furnace-brazed at the same time.

Silver soldering

This process is very similar in production and in the joint produced to brazing and is often called hard soldering. The basic difference between silver soldering and brazing is in the type of filler rod used. Silver soldering uses an alloy primarily of silver and copper (usually with more than 50% silver).

The joint is formed at a temperature lower than the melting temperatures of the materials being joined. Operating temperatures range between 600 °C and 850 °C depending on the solder being used.

The general procedure of producing the joint is the same as for brazing and can be used to join copper, brass, bronze, steel or nickel.

Self-assessment questions

1 Why has the bit of the soldering iron to be filed?

2 What is the essential requirement for achieving a good joint by soft soldering, silver soldering, or brazing?

3 What is meant by a solder 'wetting' the surface of a material?

4 Choose from the list below topics relating to soft soldering which distinguish it from brazing:

(a) Melts at high temperatures
(b) Forms a soft joint
(c) Forms a high strength joint
(d) Suitable for applications involving vibration
(e) Suitable for electrical applications
(f) Suitable for operation at low temperatures
(g) Uses a lead–tin alloy
(h) Heated with a blow-pipe

5 State the type of alloy used to form a brazed joint.

6 Why must some joints be washed after being formed?

7 What are the temperature ranges for the melting of the alloys used in soldering and brazing?

After reading the following material, the reader shall:

10.4 Describe the differences between gas and arc welding.

Welding is a form of joining two materials. It has developed into a major method of producing items by fabrication. It is also used in much repair work.

Consideration is given here to the two basic forms of fusion welding, gas and metal arc welding. In both methods heat is used to melt a welding rod, together with localized areas of the materials being joined, to form a molten pool of metal, the whole of which fuses together on solidification.

Gas welding

As the heading implies, the heat source for the process is gas. The gases which are used are acetylene and oxygen. Acetylene is used more extensively than propane.

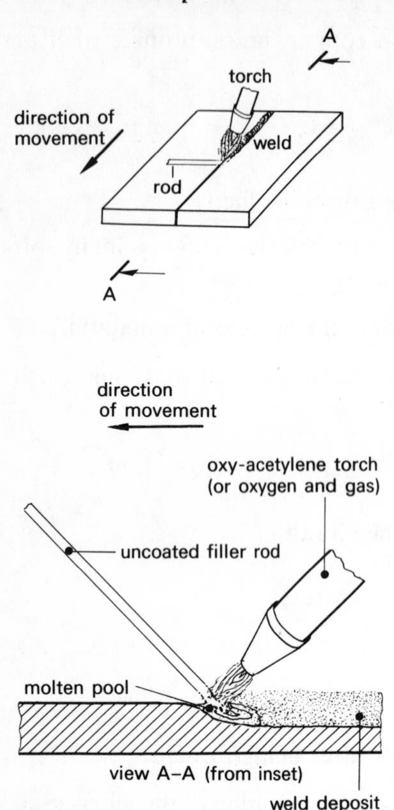

Figure 173 *Gas welding*

Using suitable mixtures of oxygen and acetylene in a blow-pipe, a flame can be obtained which is at a temperature of approximately 3500 °C. This melts most common metals and alloys very quickly. If an incorrect mixture of gases is used this is detrimental to the weld and may result in premature failure of the joint. An advantage of the metals being melted quickly is that less distortion occurs.

In the gas-welding process, the operator holds a filler rod in one hand and the welding blow-pipe in the other. The pieces to be joined are prepared and positioned ready for welding. Heat from the blow-pipe creates localized melting and the filler rod, which must be held close to the work, is melted by the heat. This material flows into the prepared joint, mixes with the molten metal from the pieces being joined forming a molten pool, and when the pool solidifies produces a strong joint.

Normally when joining similar metals the filler rod and the metals being joined are all made of the same material. The process is as shown in Figure 173.

Some form of flux is usually necessary on non-ferrous metals but not on mild steel. This dissolves the oxide film before fusion and protects the metal from oxidation afterwards.

The correct type of welding goggles must be used at all times during welding.

Metal arc welding

The metallic arc process has now developed into the most common method of fusion welding. In the process, an electric current is passed through an electrode and an arc is struck between the

Solutions to self-assessment questions

1 To clean the surface and remove dirt, oxide etc., which may prevent a joint being formed.

2 Cleanliness of the materials being joined, particularly in the joint area.

3 Atoms in the solder combine with atoms on and near the surface of the material being joined, forming an alloy.

4 The particular properties of soft solder are (b), (e), (f) and (g). The other properties refer to silver soldering or brazing.

5 An alloy of copper and zinc is usually used. The minimum content of the copper is 50%.

6 To prevent corrosion as a result of the flux used in the process.

7 For soft soldering the range is 182 – 250 °C. For silver soldering and brazing it is 600 – 900 °C.

electrode and the materials being joined. This arc creates the necessary heat to melt locally the pieces being joined. The other major difference between this and gas welding is that the electrode acts as the filler rod; it is melted, along with the localized areas of the joint, to form a molten pool which solidifies on cooling.

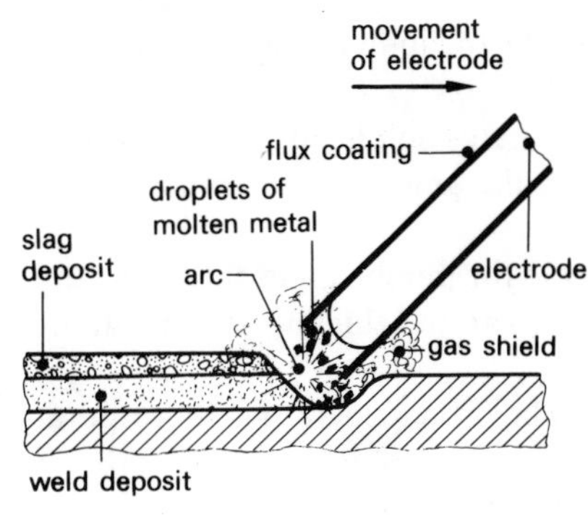

Figure 174 *Electric arc welding*

Another difference is that the electrode is normally coated with a suitable flux which provides a shield of gas around the arc, thus preventing atmospheric attack. On solidification, elements from the flux form a protective slag over the weld which must be removed after solidification. The electrode is usually of a similar type of material to the pieces being joined. The process is hand operated, the electrode being moved along the joint at an even rate as it melts, producing a continuous weld. The process is shown in Figure 174.

Steels are the most common materials joined by this process but the method is also suitable for joining other metals. An advantage of the process compared with the gas process is speed and economy, particularly with thicker metals, i.e. more than 3 mm in thickness.

One danger of the arc welding process is the flash produced when the arc is struck. This can be harmful to the eye and cause much discomfort. During the process, ultra-violet radiation is emitted, and it is essential that the correct welding shields be used at all times whilst welding is being carried out.

Self-assessment questions

8 What is the main difference in the formation of soldered and fusion-welded joints?

9 Brazing is carried out at a higher temperature than welding.
TRUE/FALSE

10 In gas and metal arc welding the joint is formed by melting localized areas of the pieces being joined together with molten metal from a filler rod. Identify three differences between gas welding and arc welding.

After reading the following material, the reader shall:

10.5 Identify the main types of adhesives and their uses.

The methods of joining described so far have been commonly used for many years – primarily for joining metals. In each instance heat is necessary to melt the flux or the filler rod or both so that the joint can be made.

Many materials are not suitable for processing in this fashion. In recent years the development of a wide range of adhesives has made the selection of jointing processes much more varied. In addition to being capable of joining parts which would be damaged by the heat, the adhesives may also be used to join metals together and to join metals to non-metals.

Nevertheless, for engineering applications, adhesives are only used when they have a distinct advantage over other methods of joining. Some of the advantages and disadvantages of adhesive bonding are summarized below:

Advantages:

(i) *Cost*. If an adhesive is suitable to make the joint it will be used as this is frequently the cheapest method.
(ii) *Neatness*. The use of adhesives eliminates bolt heads, screws, welds, or other fastenings. This improves the general appearance of the joint.
(iii) *Adaptability*. Adhesives are suitable for joining materials with very different properties. They can join varying thicknesses, including very thin sections.
(iv) *Skill*. Usually an adhesive bond can be produced by unskilled operatives.
(v) *Stress*. The stress concentrations are reduced.

In contrast to these points there are some definite disadvantages:

(i) The time involved in many instances waiting for the material to cure and form the bond. In some cases it may take a number of days for an adhesive to attain maximum strength.
(ii) There is difficulty in breaking the joint at a later stage if repairs are needed to any part of the assembly. When the joint is broken the whole area must be well cleaned before a new joint can be made.
(iii) Some adhesives give off dangerous fumes; well-ventilated areas are therefore essential.

Solutions to self-assessment questions

8 When a joint is soldered the materials being joined do not melt; in a welded joint they do. The material being used for soldering combines with atoms on or near the surface of the materials being joined to form the joint. In welding the electrode or filler rod and areas of the materials being joined adjacent to the joint, melt and form together on solidification.

9 False. Welding is carried out at a much higher temperature than soldering or brazing since the materials being joined are melted locally.

10 (i) For gas welding heating is by acetylene and oxygen gases, whilst in metal arc welding heating is produced by an electric arc.
(ii) Metal arc welding is suitable for joining sections over 3 mm thick whereas gas welding is more suitable for joining sheet metal less than 3 mm thick.
(iii) A weld produced by metal arc welding has a protective layer of slag covering it but this is not so with gas welding.

(iv) If in contact with heat or chemicals some adhesives may break down, causing the joint to fail.

(v) The strength of the joint depends upon the use of the correct adhesives. While several types may be used, some are less suitable than others. This is more important when joining plastics, since some adhesives may react with the plastic if an incorrect choice has been made.

(vi) The life of the joint when subjected to extreme environmental conditions is unknown.

As with soldering and brazing, it is vital when using adhesives to ensure that the parts are free from grease, oil, dirt, etc. When the adhesive is applied to the surfaces to be joined, it 'wets' the surface of these materials.

Self-assessment question

11 'Wetting' the surface of a material was mentioned earlier in the text. What does the term mean?

The types of adhesives can be loosely divided into four main groups.

(i) *Natural materials* including both organic and inorganic materials, the first type of glue or adhesive to be used and still in common use. These are:

(a) *Vegetable glues.* These are made from starches, dextrines, or soya beans and are used mainly in the paper and packaging industries.

(b) *Animal glues.* These are stronger glues than vegetable glues. They are produced from skins, bones or fish. They are suitable for use with wood (in furniture manufacture), in the paper industry, and for joining leather and fabrics.

(c) *Natural resins.* These include materials such as bitumins and shellac and produce adhesives suitable for felt, flooring, paper, glass and metals.

(ii) *Elastomers.* These adhesives are based on natural and synthetic rubbers. They are used for bonding papers, rubbers and nylon; and also in the footwear industry. They are generally of low strength but have high flexibility.

(iii) *Thermoplastics.* These are synthetic resins, such as vinyls, cellulose, acrylics, polystyrene and polyamides. Thermoplastic means they soften when subjected to heat. An everyday use of a thermoplastic resin is the production of a joint by bonding a layer on to a strip (e.g. edging for veneered doors). The strip is positioned on the panel and heat applied with a hot domestic iron. The heat causes the thermoplastic to soften and a bond is formed on cooling. General applications are found in joining paper, leather, wood and plastics. These resins should be used only when subjected to light loads and low temperatures.

(iv) *Thermosetts.* These are some of the strongest adhesives available.

They include materials such as phenolics, epoxy resins and polyurethanes. The adhesives set by a chemical reaction which occurs either at room temperature, or by being heated. The joints produced are strong, and are used in structural applications. The materials are resistant to water, oil, solvents and some chemicals, and are particularly suitable for joining dissimilar materials. A wide use of these materials is for bonding metals in aircraft structures where savings are made in weight, since with adhesive-bonded joints much thinner metals can be used than is normally possible with other fastening procedures.

Some glues combine two or more of the above groups. Examples of these are Evostik and Bostik which are very well known adhesives. When making a joint, the surfaces must be well prepared and free from oil, dirt, grease. This enables the bond to form.

After reading the following material, the reader shall:

10.6 Compare the relative advantages of soldering, brazing, welding and using adhesives.

10.7 Give examples of typical applications for each joining method described above.

When considering the relative advantages of each of the joining processes (soldering, brazing, welding and using adhesives), it is useful to relate them to the applications for which they are most suitable. The methods of joining described in the text range from applying no heat at all (most adhesives), through a temperature range (which includes soldering and brazing) where, with welding, the materials being joined are melted.

Strong bonds are normally produced when high temperatures are employed. Thus if all methods are applicable, the deciding factor must be the required strength. The strongest bond is a welded joint. At the opposite end, the adhesives generally produce the lower-duty joints, although structural load-bearing joints can now be produced which are used in the aircraft, automotive, marine and civil-engineering industries.

In the middle range are soldering and brazing. These processes are ideally suitable for joints where the temperatures involved may be too high for adhesives. They are also suitable where the high heat of welding may damage the structure of the metal.

Solution to self-assessment question

11 Atoms in the adhesive combine or are attracted to atoms on or near the surface of the material being joined. This is the method by which a soldered or brazed joint is formed.

Welding is restricted to metals (except for the special methods used with plastics), but adhesives are suitable for almost any material. The welding, soldering and brazing processes require much more skill to produce joints than the processes using adhesives.

Typical applications are:

Soft soldering: electrical wiring and connections, general plumbing work, tinplate (including tin cans), light-duty low-temperature applications (such as car radiators).
Silver soldering and brazing: higher-duty engineering applications, fixing tips to cutting tools, joining brass, copper and stainless steels, joining of iron and steel.
Welding: general fabrications of all sizes requiring strong continuous joints. Typical examples are in the automobile industry, ships structures, rolling mills, structural steel-work, brackets, tanks and pressure vessels, pipework, etc. In addition the arc process and gas welding can be used for coating parts of a component with a special surface, e.g. a hard-wearing surface. The apparatus needed for the welding processes is portable and is used extensively in repair work.
Adhesives: furniture, plywoods, laminates, seals for cars and in the home, general duty around home, etc. Production of models, floor coverings, paper envelopes, marine, air-craft and electronic industries, footwear manufacture and packaging applications.

Self-assessment questions

12 Why must the joint be clean for an adhesive to bond?

13 The bond formed by a thermoplastic withstands heat.
TRUE/FALSE

14 The following adhesives have varying strengths: starch, synthetic rubber, and phenolic. Place them in order of strength with the highest strength first.

15 What purposes does the flux perform in metal arc welding?

16 Which welding process requires the use of both hands?

17 For each of the examples below suggest a suitable method of joining the assembly:
(a) Domestic cold water piping (copper)
(b) Steel framework for a support panel
(c) Fixing the wooden legs on dining chairs
(d) Fixing wall tiles
(e) Joining two pieces of similar plastic
(f) Re-tipping a lathe tool

18 State the main difference between joints produced by soft soldering and joints produced by brazing.

Solutions to self-assessment questions

12 It bonds by 'wetting' the surfaces to be joined. It cannot do this if the joint is dirty. This is similar to soldering and brazing.

13 False. A thermoplastic may soften when subjected to heat.

14 1st phenolic; 2nd rubber; 3rd starch.

15 Some of the elements provide a gaseous shield whilst others form into a slag on top of the weld, protecting it from atmospheric attack.

16 Gas welding. One hand holds the filler rod whilst the other hand holds the welding blow-pipe.

17 (a) Soldering
(b) Welding – arc
(c) Adhesive
(d) Adhesive
(e) Adhesive – may be welded with special process
(f) Brazing

18 Soft soldering forms a relatively weak joint which only withstands low stresses. Brazing produces stronger joints.

Topic area: Materials

After reading the following material, the reader shall:

11 Know what is meant by the term 'material'.
11.1 Outline the fact that materials are made up of matter and everything we see and use is a 'material'.
11.2 Identify some of the most common materials used by engineers and technologists.
11.3 Explain that the properties which materials possess enable the material to perform specific duties, and each material has its own exclusive range of properties.

12 Know the properties of materials.
12.1 Distinguish between physical and mechanical properties.
12.2 Describe physical properties of materials, thermal conductivity, electrical conductivity, melting point, density.
12.3 Describe mechanical properties of materials, ductility, malleability, toughness, strength, brittleness, hardness and elasticity.
12.4 Compare the cost, strength, thermal conductivity, electrical resistivity, density, melting point and elongation of common plastic and metallic materials.

Engineers and technologists require a large range of materials to fulfil the requirements and performances of designs. These materials may be subjected to extremes of conditions so that no one material is adequate for all purposes.

Everything that can be seen, everything that can be touched or used is a material. All are made up of numerous atoms arranged and bonded in specific ways to create the materials as they are known. These arrangements of atoms and bonds are the fundamental basis of the properties of the material, properties which can be modified, within limitations, to enable the material to perform a wider range of duties.

Self-assessment question

1 Make a list of materials that are commonly used in engineering.

The materials listed are just a few of the range available, each having its own individual properties. Whilst all materials have individual properties they all have some which are common and basic. These properties can be divided into two categories:

(i) Physical properties
(ii) Mechanical properties

Self-assessment question

2 What is the difference between the two types of properties?

Physical properties

Many properties fall into this category, but this section is restricted to the ones most relevant to engineers.

Thermal conductivity

Heat travels from the hot end of the material to the cold end by vibration of atoms in the structure. The heat energy transferred to a material causes the atoms in the material to increase the rate at which they vibrate about their fixed point in the lattice. When the material is a metal, electrons in the outer shells of metal atoms readily escape from their parent atoms. These electrons are called free electrons and become readily available to transfer energy to other atoms within the material. The number of free electrons increases as the temperature is raised. The external effect is that heat energy is transferred through the material from the high-temperature zone to the low-temperature zone. The heat flow rate is in watts (W) and the thermal conductivity (λ, the Greek letter 'lambda') is in watts per metre kelvin (W/mK). The thermal conductivity of metals is higher than the thermal conductivity of materials like wood, asbestos, paper and felt.

Electric conductivity

All substances are made up of atoms. All atoms have a nucleus containing a positive charge. The positive charge is on the protons in the nucleus. The positive charge on a proton is exactly balanced by the negative charge on an electron. The electrons orbit the nucleus. The electron orbits arrange themselves in definite regions around the nucleus. These regions are called shells. The maximum number

Solutions to self-assessment question

1 Typical materials are: low-, medium- and high-carbon steels, copper, brass, bronzes, aluminium, cast iron; nylon, polythene, polypropylene, bakelite, PVC.

of electron orbits in any one shell is fixed. The maximum number of electron orbits in the outer shell of any atom is eight. When the outer shell contains its maximum of eight electrons orbits the atom cannot readily accept or lose an electron. When the outer shell is not complete the atom can readily accept or donate electrons.

An electric current is a unidirectional drift of electrons. In all solid materials the charged particles which are the electric current are electrons.

Solid materials in which the outer shells of the atoms which make up the material are complete have few if any electrons available to produce a unidirectional drift of electrons. These materials, e.g. wood, PVC, glass, paper, ceramics etc., have low electric conductivity (and low heat conductivity). Solid materials in which the outer shells of the atoms which make up the material are incomplete readily lose or gain electrons, i.e. they contain electrons which move in a random direction from one atom to another in the material. These free electrons are available to form a unidirectional drift of electrons, i.e. a current flow. These materials have a high electrical conductivity (and high thermal conductivity).

To produce a unidirectional drift of electrons through a substance a difference in the magnitude of the state of charge between the ends of the substance is required, i.e. one end of the substance must be more negatively charged than the other. The difference in charge produces forces of attraction and repulsion on the negatively charged electrons in random motion in the material. These forces order the random motion into a unidirectional drift. The unidirectional electron drift is opposed by electrons in the drift and by those atoms which have not a negative charge. The opposition to the electron drift is called resistance (R) and is given the name ohm (Ω). The unidirectional drift is assisted by those atoms which have lost an electron and those have a positive charge which attracts the negatively charged electrons. This assistance to the electron drift is called conductance (C). The unit of conductance is siemens (S). This unit was formerly the mho (the opposite of ohm). The property of a material not to oppose/assist a current flow is called electrical conductivity (σ). It is measured in kilosiemens per millimetre (kS/mm). Note that high conductivity is the equivalent of low resistance, low conductivity is the equivalent of high resistance.

Materials with low conductivity are classed as insulators, i.e. materials which oppose the flow of current. Examples include ceramics, plastics, wood, paper and vegetable oils (switch oils). Materials which have high conductivity are classed as conductors, i.e. they offer little opposition to the flow of current. Examples include all metals – platinum, gold, silver, copper, lead, iron and steel, etc.

All low-conductivity materials become conductors under extremes of temperature and high potential differences – consider the effect of lightning on air, wood, stone, ceramics and PVC.

Melting point

This is the point where the last of the bonds between the atoms in a solid material finally breaks. When all the bonds are broken the material is completely liquid.

Engineers need to know this temperature for a number of important reasons:

(a) For casting the material into moulds
(b) For making alloys with other materials
(c) For hot working in the plastic stage

The melting points of materials are very specific points (Figure 175). When two or more materials are combined, as in alloys, then the resulting melting point usually varies depending upon the proportion of each material being added.

metal	*melting point (°C)*
aluminium	660
copper	1083
gold	1063
iron	1535
lead	327
manganese	1244
mercury	−39
nickel	1453
tin	232
zinc	419

Figure 175 *Typical melting points*

Density

Density is defined as 'the amount of matter or substance contained in unit volume of material'.

$\therefore$ *density* = *mass* per *unit volume*

or $\text{density} = \dfrac{\text{mass}}{\text{volume}}$

Now consider the units: always work in basic units –

mass – units are kilograms (kg)
volume – units for length are metres (m)
$\therefore$ units of volume are metres × metres × metres = metres^3 (m^3)

$\therefore$ The units of *density* are $\dfrac{\text{kg}}{\text{m}^3}$

Consider the following example:

A block of material measuring 100 mm × 100 mm × 500 mm has a mass of 2000 gm. What is the density of the material?

Solution to self-assessment question

2 Physical properties are the properties inherent in the material due to the chemical and physical structure of the matter in the material.

Mechanical properties are the properties produced by the material whilst being subjected to various load conditions.

Convert to basic units:

$$\text{volume} = \frac{100}{1000} \times \frac{100}{1000} \times \frac{500}{1000} \text{ m}^3$$
$$= 0.1 \times 0.1 \times 0.5 \text{ m}^3$$
$$= 0.005 \text{ m}^3$$
$$\text{mass} = \frac{2000}{1000} \text{ kg}$$
$$= 2 \text{ kg}$$
$$\text{Now density} = \frac{\text{mass}}{\text{volume}} = \frac{2}{0.005} \frac{\text{kg}}{\text{m}^3}$$
$$= 400 \frac{\text{kg}}{\text{m}^3}$$

There are other properties such as boiling point, specific heat, coefficient of thermal expansion, to name a few vital to designs by engineers and technologists. The importance of each of these properties depends upon the application and conditions to which the material is subjected.

Mechanical properties

These properties are very important to the engineer; they influence both the design and the choice of production procedure. Each of the properties can be modified, which may change the design or the method of production. No material can exhibit all the properties but each has its own specific range or combination. It is this combination of properties which the engineer must consider when deciding what the component will be made from in relation to the duties the component has to perform.

Self-assessment question

3 List some of the main factors in the choice of materials for engineering components.

These are some of the important considerations which the engineer or designer must take into account. Many ideas are gained from long experience and application either by the engineer himself or passed on from other sources.

The basic mechanical properties of metals are:

1 Ductility
2 Malleability
3 Toughness
4 Strength
5 Brittleness
6 Hardness
7 Elasticity

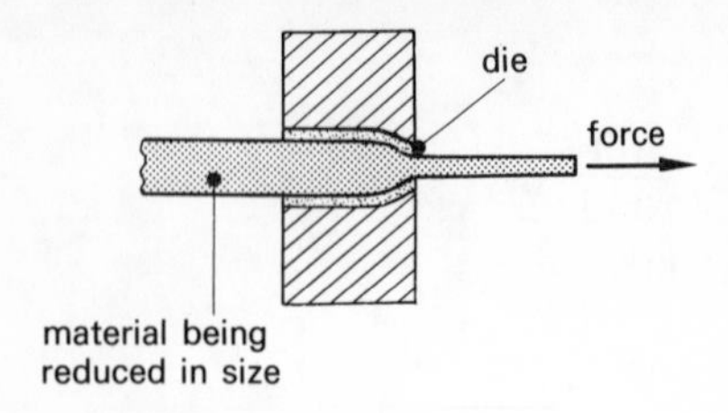

Figure 176 *Diagrammatic representation of wire drawing*

Ductility

A ductile material has the ability to 'flow' and be drawn out when subjected to tensile loading. This means that the material can undergo plastic (permanent) deformation without fracture. When this occurs some atoms break their bonds, move and form new bonds. The result of the deformation is a change of shape of the material, the extent of which depends upon the loads applied.

Ductility determines the amount that a metal can be cold-worked without fracture. If the metal is heated the ductility is reduced due to the lowering of the material strength as the temperature increases.

A typical example of a manufacturing process which is dependent upon the ductility is the drawing of wire (Figure 176).

Malleability

A malleable material has the ability to withstand plastic deformation (without fracture) when subjected to compressive forces. Malleability and ductility are similar properties and some materials possess both. However, it must not be assumed that if a material has one of the properties it has the other: malleable materials tend to be weak in tension and may tear when loaded.

The malleability of a material is normally increased when the material is subjected to higher temperatures; use is made of this during some primary forming processes. Rolling, forging, extruding and pressing are examples of processes requiring malleability (Figure 177).

Toughness

This property enables a material to withstand bending and forming without fracture. A tough material also withstands shock loading.

All materials contain cracks, notches and other imperfections. Many of them are invisible to the naked eye but all are possible sources of premature failure.

Solution to self-assessment question

3 strength — operating temperature
toughness — working conditions of component
hardness — shape – can it be produced?
wear
fatigue — appearance

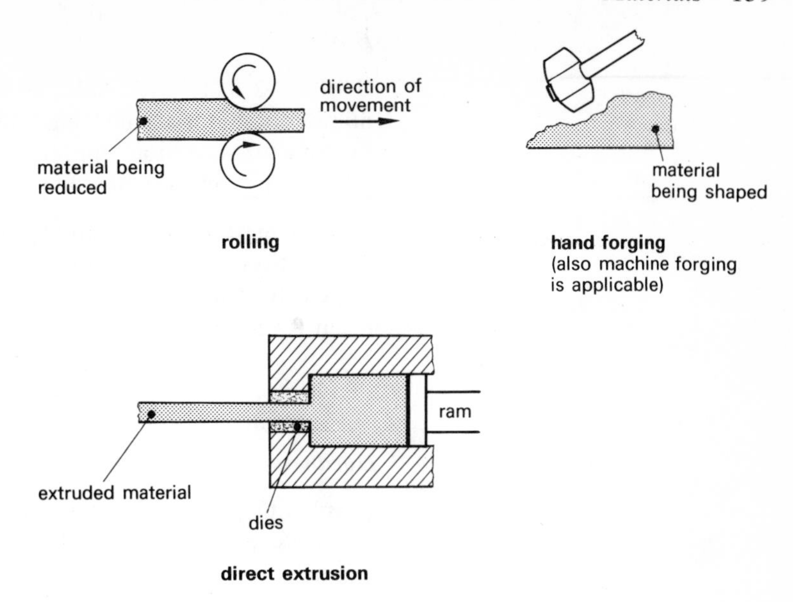

Figure 177 *Processes requiring malleability*

The forces that are applied to the materials tend to increase the size of these cracks and defects until a fracture occurs. Toughness is the resistance of the material to the growth of these cracks and defects; the more resistance, the tougher is the material.

Strength

This property is of interest to all who use materials. Here, engineers are considering the possible breaking of a material or component. No material is indestructible or unbreakable, but the amount of force or stress which is required to tear apart the bonds between the atoms and break the material varies from material to material and also depends upon the type of the force or stress.

The strength of a material can be defined as the 'ability to withstand the application of forces without fracture'.

A number of different forms of strength may be considered during design: tensile strength, compressive strength, and shear strength. The different strengths are related to the type of forces applied to the material. The strength most commonly used is tensile, particularly for structural materials; the tensile strengths are usually quoted in engineers' handbooks (Figure 178). Tensile strength is measured as the maximum load necessary to fracture unit area of a material.

material	*strength* (MN/m^2)
steel (mild)	460
aluminium	80
copper	150
lead	15
tin	30
zinc	150

Figure 178 *Typical tensile strengths*

The units of strength are MN/m^2.

Brittleness

Brittleness can be thought of as the opposite of toughness. A number of materials have high theoretical strengths but are brittle, i.e. they break without appreciable deformations and at low stress. Glass is a good example.

A tough material has the ability to resist, within limits, the growth of cracks. In contrast, a brittle material offers little resistance, breaking easily and quickly. The energy required for crack growth to occur is very low and the crack grows rapidly through the material.

Hardness

Possibly the best way to consider hardness is by looking at the way hardness readings are obtained. There are two primary forms of testing:

(i) Indentation of the material by either a hardened steel ball or a diamond
(ii) Scratching the surface of the material

In the first series of tests the balls or diamonds are forced by known loads into the surface of the material. The size of the indentation is measured and this is related on a chart to a hardness number. The most commonly used methods are:

(a) The Rockwell Hardness Test – using either balls or diamonds with a range of loadings, each having related scales
(b) The Brinell Test – using a ball with a range of loads
(c) The Vickers Pyramid Test – using a diamond

Each test therefore has its own scale of hardness which can be related and compared.

The Vickers method is most reliable because of the accuracy of the machine and the use of a diamond, which deforms far less when subjected to heavy load than a ball.

The second type of testing, the scratch test, is used for materials which would normally break if subjected to the compressive loads used in the Rockwell, Brinell or Vickers tests.

Scratch testing is more applicable to materials such as plastics and ceramics. It is rarely used in comparison to the indenters.

The property of hardness can therefore be summarized as the resistance that a material offers to deformation or scratching. The smaller the deformation or the more difficult to mark the surface by a scratch, then the harder is the material.

Elasticity

When a material is subjected to forces it deforms in some way. The size of the deformation varies, ranging from very small, almost undetectable amounts, to considerable amounts seen clearly by just a glance. A number of factors contribute to the size and type of deformation.

Some of these factors are:
(a) Magnitude of applied forces
(b) Nature of the forces (tension, compression, shear, etc.)
(c) Method of application of forces
(d) Condition and type of material

One important factor is the nature of the material itself. Each material behaves in its own way due to structure and bonding of the atoms, etc.

There are two main types of deformation. One is temporary (elastic), and the other is permanent (plastic).

This book is interested in the elastic or temporary deformation. During this stage the material deforms due to the applied forces, but returns to its original shape and size when the forces are removed. This is an in-built property of the material which depends upon the structure of the material and the strength of its bonding.

material	E (GN/m^2)
steel	210
aluminium	71
copper	117
lead	18
tin	40
zinc	110

Figure 179 *Typical values for Young's modulus of elasticity (E)*

The elasticity of the material can be said to be the resistance offered by the material to elastic or temporary deformation when subjected to forces. This can also be regarded as a measure of the stiffness of the material. The measure of elasticity is given the name 'Young's modulus of elasticity', and is a constant for each material. The modulus figure is quoted extensively in reference books and used in designs (Figure 179).

Self-assessment questions

4 Choose one answer from (a), (b), (c) and (d) to satisfy the following question.

If a material has the property to resist the growth of cracks it is said to be:
(a) strong
(b) ductile
(c) tough
(d) hard

5 Define brittleness.

6 What is the difference between strength and toughness?

7 A malleable material has the ability to withstand deformation plastically when subject to tensile loading.

TRUE/FALSE

Comparison of properties

Costs of components are very difficult to compare for a number of reasons:

(i) Any figures quoted are only valid at the time of printing, and the price variation is not usually a linear relationship between all materials.

(ii) If the density of the component is not considered then a true relationship of what is being bought is not obtained e.g. plastic occupies a very large volume compared with the same weight of steel.

(iii) Due to differences in strength and density etc., a component which can be manufactured in several materials, e.g. steel, PVC or aluminium, must be designed in a specific way for each material to allow for its own properties. This means that comparisons would be made of incompatible items.

(iv) Manufacturing procedures are vastly changed for producing components in different materials.

Due to these problems actual cost figures are not to be included in the following text. The other comparisons are tabulated to give easy correlation (Figure 180).

After reading this section the reader shall:

13 Know the compositions, properties and recognize the uses of some common ferrous materials.

13.1 State the difference in composition and properties of low-, medium- and high-carbon steels and cast iron.

13.2 State the individual stages of heat treatment processes, annealing, normalizing, hardening and tempering, as applied to simple carbon steels.

13.3 State the modifications to the properties of steel produced by heat treatment.

13.4 Identify the treatment necessary to produce necessary modifications to properties of plain carbon steels.

13.5 Relate the materials/properties from 13.1 to typical engineering applications.

Perhaps the most common and widely used material is steel. It is produced in a wide range of compositions, which allows the engineer great flexibility in design.

material	*density, ρ* (kg m^{-3})	*tensile strength* (MN/m^2)	*thermal conductivity* (Wm^{-1}K^{-1})	*electrical resistivity* × 10^{-8} (Ωm)	*melting point* (°C)	*elongation* (%)
plain carbon steel (low)	7 860	460	63	15	1427	35
grey cast iron	7 150	100	75	10	1227	
white cast iron	7 700	230	75	10	1147	
aluminium	2 710	80	201	2.65	660	43
brass (70%Cu/30%Zn)	8 500	550	110	8	1027	8
copper	8 930	150	385	1.7	1083	45
lead	11 340	15	35	21	327	50
tin	7 300	30	65	11	232	
nylon	1 150	70	0.25		197	60–300
polyethylene (high density)	955	26			137	100–300
polypropylene	900	35	0.08		177	> 220
PVC (rigid)	1 700	60			212	5–25
phenolformaldehyde	1 300	50	0.2			0.4–0.8
perspex	1 190	50	0.2		77	2–7

Figure 180 *Comparative properties of materials*

Steel is basically an alloy of iron and carbon, having a carbon content of up to approximately 1.7%. When steels consist primarily of iron and carbon they are termed plain carbon steels. In practice there are also small amounts of other elements included which are there either due to the manufacturing process or purposely added to perform specific duties. Elements normally included in plain carbon steels are silicon, phosphorus, manganese and sulphur. However, these additives are neglected in the following text and the steel is regarded as being iron and carbon – the combination of these two elements is by far the most important factor for consideration.

Plain carbon steels

Plain carbon steels are divided into three main categories – low-, medium- and high-carbon steels.

Solutions to self-assessment questions

4 (c) tough

5 The tendency to fracture without appreciable deformation and at low stress.

6 Strength is the ability to resist loading or forces without fracture.

Toughness is the ability to withstand bending and forming without fracture – a resistance to crack growth.

These steels, as would be expected, are mainly distinguished by their carbon content. The division is approximately:
low-carbon steel up to 0.30% carbon
medium-carbon steel 0.30 – 0.60% carbon
high-carbon steel 0.60 – 1.4% carbon

Whilst steel may have a carbon content up to 1.7%, it is very rare that more than 1.4% is used.

Low-carbon steels

The lower the quantity of carbon, the more the material behaves like pure iron. The hardness of a steel is increased by the addition of the carbon; since the proportion of carbon is low the hardness figure is low. These materials normally have good ductility and are easily worked. Steels with carbon content between 0.15 and 0.30% are called mild steel. This is a very common steel extensively used throughout industry.

The list below contains some of the many uses/applications of low-carbon steels and these have been divided into specific carbon content for clarity:

carbon content	*application*
up to 0.15%	Rivets, wire, nails, chain, seam-welded pipes, stampings, hot- and cold-rolled strip – which is used in many other applications
0.12 – 0.20%	Structural steel sections, machined parts, sheets and plates, drop forgings and stampings, fabrications
0.20 – 0.30%	Structural steelwork, levers, forgings, shafts

Medium-carbon steels

This range of steels has a greater carbon content than low carbon steel, i.e. between 0.3 and 0.6%.

Self-assessment question

8 What effect does increased carbon content have on the hardness of steel?

Solution to self-assessment question

7 False.

Since the content of the carbon is increased, the hardness of the steel also increases. This increase means that the material is less ductile and a little more difficult to work. Whilst the material loses some of its ductility it is offset by an increase in strength. This increase in strength can therefore be used to perform duties of which the lower-carbon steel is not capable.

Typical uses of medium-carbon steels are:

carbon content	*application*
0.30 – 0.42%	shafts (higher duty), forgings, connecting rods, axles, crane hooks
0.40 – 0.50%	Shafts, gears, axles, machined parts requiring heat treatment, crankshafts
0.50 – 0.60%	Rails, wire ropes, leaf springs

High-carbon steels

These steels are obviously the high-duty steels. The carbon content, whilst varying considerably throughout the range, is very high. Since the carbon content is very high the steel is much harder, and further degrees of hardness can be achieved by heat treatment. A steel from this category is used mainly where advantage can be obtained from the extra hardness. One use for this particular property is for tools and this range of compositions is used extensively in the manufacture of hand and machine cutting tools.

Typical uses of high-carbon steels are:

carbon content	*application*
0.60 – 0.80%	dies, set screws, saws, anvil faces, hammers, laminated springs
0.80 – 0.90%	chisels, punches, hand tools, shear blades
0.90 – 1.00%	knives, springs, dies, picks
1.00 – 1.10%	drills and taps, milling cutters
1.10 – 1.40%	lathe tools, boring tools, files, razors, reamers, wear-resistant parts

The applications listed for all carbon steels are only a selection of the more important uses.

All steels have their own properties; an additional range of properties can be achieved in most cases by heat treatment. The type of heat treatment used influences the final property of the steel.

Self-assessment questions

9 A component which is to be manufactured from a plain carbon steel requires a composition with maximum ductility. Which category would be most suitable?

10 State whether low-, medium- or high-carbon steel should be used for the following applications.

(a) heat-treated pin
(b) welded bracket
(c) guillotine blades
(d) forgings

11 Why are cutting tools manufactured from high-carbon steel in preference to low- or medium-carbon?

All metals are polycrystalline. This means that their structure consists of many similar crystals (or grains). These crystals are regular three-dimensional structures of atoms held in specific positions by interatomic forces. The total symmetry of the structure is spoiled by randomly positioned defects, of varying types, which occur during solidification. It is these defects which affect and/or influence the basic material properties by their positioning, or their ability to move through the structure.

Heat treatment is therefore a process which modifies the basic structure of a material to produce new properties. There are many forms of heat treatment but emphasis will only be placed on the most common and widely used. The treatments which are used most extensively, and are suitable for most materials, are annealing, normalizing, hardening and tempering. Individual treatments must be considered for individual materials, because some materials react in similar ways whilst others can react entirely differently, producing extremes of properties.

Annealing of plain carbon steels

Annealing is a process used mainly for materials which have undergone cold-working. During the working the material becomes highly stressed as its structure is distorted. Apart from being highly stressed, the material has become harder. For satisfactory use and prevention of premature failure these stresses must be reduced and

Solution to self-assessment question

8 The hardness is increased by the carbon content.

the hardness reduced. This is achieved by annealing. Annealing softens a material and at the same time recrystallizes the structure, thus returning it to a similar structure to the one possessed by the material before any work was carried out on it.

There are three basic stages in annealing:

1 Stress relief
2 Recrystallization
3 Grain growth

Stress relief takes place in steels at temperatures up to approximately 200 °C. No change in structure occurs during this stage and light relief only is achieved.

Recrystallization: as the temperature is increased a point is reached where nuclei form, usually at high stress points, and these form new crystals. The new crystals grow gradually at the expense of the distorted crystals until the whole of the structure is transformed.

The final stage is *grain growth*, which occurs if the temperature is allowed to increase after recrystallization has taken place. Some of the crystals continue to grow at the expense of others in the structure. This grain growth must be controlled; if it is not the material becomes very brittle and may fail prematurely. Annealing temperature ranges for plain carbon steels are shown in Figure 181. An important stage in annealing is the rate of cooling. Once the material has been recrystallized into crystals of a reasonably uniform size and shape, the crystals must be maintained by very slow cooling. This is normally carried out by switching off the furnace and allowing furnace and material to cool together.

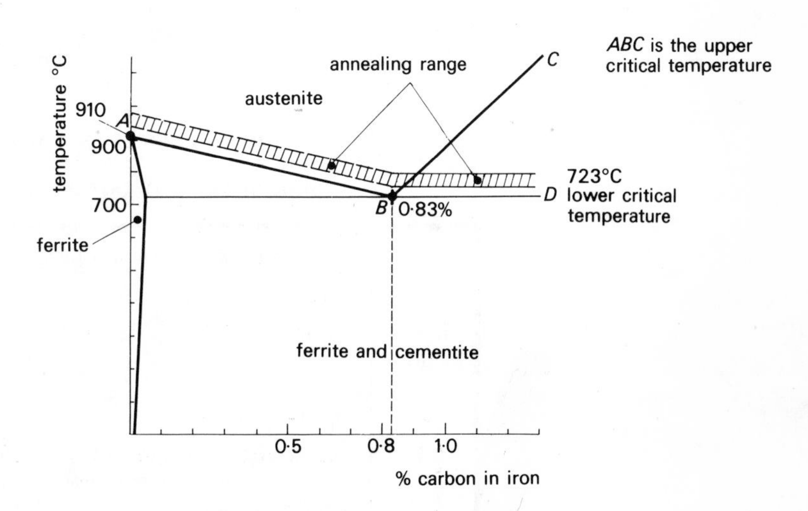

Steels up to 0.83% carbon are heated to between 30 °C and 50 °C above the upper critical temperature. Steels above 0.83% carbon are heated to between 30 °C and 50 °C above the lower critical temperature.

Figure 181 *Annealing temperatures for plain carbon steels*

Figure 181 shows part of the steel section of the iron-carbon phase diagram. These diagrams are very important but very complex so they will not be discussed in detail here. The line ABC on the diagram represents the line called 'the upper critical temperature'. Above this line the structures of the materials become uniform crystals again. This uniform structure is essential to achieve the desired properties after annealing and therefore during the treatment the materials are heated as shown in Figure 181, and then slowly cooled.

Changes in the structure of steels during annealing are shown in Figure 182.

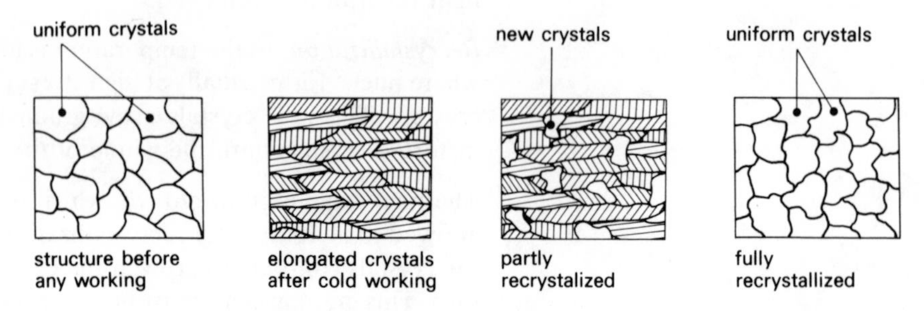

Figure 182 *Changes in the structure of steels. Recrystallization occurs in steels at approximately 550 °C*

After annealing, the steels are in their softest and most ductile condition. They are also in a condition to undergo further cold working if this is necessary for achieving specifications.

Annealing is usually carried out after the whole of the cold working is complete and also at intermediate stages if the working is extensive.

Due to the very slow rate of cooling the crystals or grains are uniform in shape and size but, most important, they are large.

Normalizing of plain carbon steels

Normalizing is a process similar to annealing. Its main object is to

Solutions to self-assessment questions

9 Low-carbon steel.

10 (a) Medium
(b) Low
(c) High
(d) Medium or low

11 High-carbon steels contain more carbon and are therefore harder both initially and after heat treatment. In most instances the cutting tool must be harder than the material being cut.

produce a uniform grain (since 'grain' is used more widely by engineers and metallurgists, it will be used in the text instead of 'crystal') throughout the structure of the material, and at the same time to relieve any high stresses created during working. As with annealing, the steel is heated up to between 30 °C and 50 °C above the upper critical temperature.

Self-assessment question

12 What are the three stages the material goes through during this heating?

The first part of the process is exactly as for annealing. When the normalizing temperature is reached it is held at that temperature for a sufficient length of time to allow the heat to soak through the whole of the material and so obtain uniform grains. The steel is then extracted from the furnace and allowed to cool in still air. This is a faster rate of cooling than annealing and maintains the fine or small uniform grains obtained during heating.

The grains are refined. Since the grain sizes are different in annealing and normalizing, the properties obtained are different. The smaller grain obtained during normalizing gives a small increase in strength but ductility is reduced compared with the annealed structure. Steels are also a little harder when normalized than when annealed.

Hardening of plain carbon steels

When a material is hardened it is made more resistant to indentation or scratching. The hardening process consists of heating the material (in this case steel) to a specific temperature and then cooling rapidly (quenching) in some medium. The quenching media most commonly used are oil and water, cold water quenching the steel more rapidly than oil. The correct type of quenching medium for each material must be used because cracks may occur if quenching is too rapid. This rapid quenching of the steel prevents the atoms having the time they need to combine in their usual bond positions, with the result that the uniform structure normally formed on solidification becomes distorted. This hinders movement within the structure causing hardness. In steel it is the carbon atoms which cause the distortion; since the content of carbon in low-carbon steels is small there are insufficient to cause any worthwhile hardening. The process of hardening therefore is restricted to medium- and high-carbon steels.

For steels below 0.83% carbon the temperatures to which they are heated before quenching are the same as for annealing. Above

0.83% carbon the temperatures used are between 750 °C and 770 °C.

When a carbon steel is fully hardened it is brittle and contains high stresses set up during quenching. Unless extreme hardness is required for a specific duty it is advisable to subject the steel to further heat treatment to reduce brittleness and stresses and achieve a vastly improved range of properties. These properties can only be achieved first by hardening and then by subsequent heat treatment. The subsequent heat treatment is called tempering.

Tempering of plain carbon steels

The desired properties of the steels are frequently attained by tempering. The temperature for tempering can range from 200 °C to 600 °C.

During the initial heating of the steel, up to 200 °C, stresses set up during hardening are relieved. No change occurs in the structure. Between 200 °C and 400 °C changes do occur in the structure of the steel and this temperature rise also causes a change in the properties. Tensile strength (which is high after hardening) and also hardness slowly reduce as the temperature increases. In contrast, ductility and toughness are very low in hardened steel and these properties tend to increase during the heating. Steels tempered within this range are suitable for applications requiring hardness and strength, typical uses being the high-carbon steels from which tools are manufactured, the tempering temperature usually being in the region of 200–300 °C.

Above 400 °C further changes occur in the structure; the important facts are that strength and hardness continue to reduce whilst ductility and toughness increase rapidly. This treatment is used for stress-bearing components which require toughness together with a reasonable strength. The components generally have good reliability and most of the steel components subjected to this treatment are medium-carbon steels.

During the treatment the steels are allowed to remain at the required tempering temperature to ensure the temperature is even throughout the structure and finally cooled slowly to room temperature.

The design of the component is influenced by the conditions in

Solution to self-assessment question

12 Stress relief, recrystallization and grain growth.

which it operates, the combination of the two guiding both the choice of material and the heat-treatment procedure.

Self-assessment question

13 Which of the following processes is used to soften fully a work-hardened steel?

(a) Tempering
(b) Normalizing
(c) Hardening
(d) Annealing

14 As the tempering temperature of a fully hardened steel increases, ductility of the material decreases.

TRUE/FALSE

15 Which heat treatment is required to reduce brittleness after hardening?

Cast iron

Cast iron is another widely used material and is also produced from iron and carbon. The major differences between cast iron and plain carbon steels are:

(a) The carbon content: cast iron contains between 2% and 4%.
(b) The form that the carbon takes within the structure of the material. In cast iron this is usually as flakes of graphite or as brittle iron carbide, cementite. Each form creates different properties.

As with steel, other elements are added in small quantities. The elements, apart from carbon, most commonly added are silicon (1.0–3.0%), manganese (0.5–1.0%), sulphur (up to 0.1%), and phosphorus (up to 1.0%). Sulphur tends to stabilize cementite, causing excessive brittleness, and is therefore kept to a minimum. Manganese toughens and strengthens an iron. Phosphorus, whilst forming a brittle compound, improves the fluidity of the material and gives better casting qualities to the iron – particularly where thin sections are to be cast and mechanical properties are unimportant.

Silicon is added to cast iron in quantities up to 3% to help form graphite instead of cementite.

Cast irons are usually classified as either ‘white’ or ‘grey’ irons.

White cast irons are given the name because the appearance of a fractured surface is bright and silvery. This is caused by the carbon taking the form of cementite. The cementite causes the iron to be very hard with good strength but there is a tendency to brittleness. Due to its hardness white cast iron is very difficult to machine.

Usually this is done by grinding, and the uses of white cast iron are very limited. Its most common applications are grinding mill parts, crusher equipment, abrasion-resisting parts and rolls for the cement, ceramic and mining industries.

The 'white' irons are used as a basis when producing malleable irons.

In *grey cast irons* the high silicon content causes the cementite to change form into flakes of graphite which give a fracture surface a dull grey appearance. The resulting material is softer with a lower tensile strength. However, in contrast, grey irons are very strong in compression and also absorb vibrations. Since grey cast iron is weak in tension it is used almost exclusively in compressive loading. The carbon content is usually between 3% and 4%, and this range has a reasonably low melting point with good fluidity giving good casting properties. Irons in this range have good machinability which makes them suitable, in conjunction with their other properties, for many uses.

Typical applications are cylinder blocks for cars, machine tool beds, wheel drums (brake drums etc.), industrial furnace parts, water pipes, pistons.

Grey irons sometimes have areas which are chilled (cooled quickly) which results in the localized area being 'white' iron. This gives a component a local wear-resistant surface whilst retaining its basic properties as grey iron.

The cooling rate is critical for cast irons. The slower the cooling the more graphite is formed; fast rates produce cementite. The correct combination of silicon content and cooling rate produces very fine graphite which results in a tough and strong material.

Cast iron is an important material for the following reasons:

(i) When molten it has good fluidity, producing good component features.
(ii) It machines easily when suitable compositions are used (grey irons).
(iii) It is rigid and has very good compressive strengths.
(iv) The material is relatively cheap, as it is produced by small modifications to the compositions of pig iron.
(v) It is suitable for further production treatments, to produce high-duty irons which have much improved properties.

Solutions to self-assessment questions

13 (d) annealing

14 False. The higher tempering temperatures normally increase ductility.

15 Tempering.

Self-assessment questions

16 Cast irons are better in tension than plain carbon steels.
TRUE/FALSE

17 What is the basic difference in the compositions of cast iron and plain carbon steel?

Additional assessment questions – properties of materials

18 Suggest whether low-, medium- or high-carbon steel should be used for the following applications:
(a) nails
(b) ball bearings
(c) car bodies
(d) crankshafts
(e) heat-treatable machine parts
(f) drills

19 Select from the list which applications are more suitable for 'white' or 'grey' irons:
(a) Machine tool beds
(b) Crusher hammers
(c) Rollers in cement works
(d) Brake drums
(e) Water pipes
(f) Furnace parts

20 What are the two forms which carbon can take in cast iron and which type of iron does it produce?

21 List the effects of tempering on strength, hardness and toughness of plain carbon steels.

22 State whether the carbon contents below relate to low-, medium- or high-carbon steels:
(a) 0.29%
(b) 0.65%
(c) 0.87%
(d) 0.08%
(e) 1.32%
(f) 0.31%
(g) 0.59%

23 What effect does annealing have on the properties of a steel?

24 In which ways does the normalizing treatment differ from the annealing treatment?

25 A steel has less strength after normalizing than after annealing.
TRUE/FALSE

Solutions to self-assessment questions

16 False. Cast irons are better in compression.

17 Both are produced primarily from iron and carbon. The carbon content of carbon steels is up to 1.7%. For cast iron it varies between 2.0% and 4.0%.

Solutions to additional assessment questions

18 (a) Low
(b) High
(c) Low
(d) Medium
(e) Medium
(f) High

19 (a) Grey
(b) White
(c) White
(d) Grey
(e) Grey
(f) Grey

20 Cementite – producing 'white' iron
Graphite – producing 'grey' iron

21 As the tempering temperature increases both strength and hardness reduce. However, in contrast to these toughness increases.

22 (a) Low
(b) High
(c) High
(d) Low
(e) High
(f) Medium
(g) Medium

23 Annealing is used for softening a material and produces maximum ductility.

24 The only difference between annealing and normalizing treatments is the way in which they are cooled. Material which is being annealed is cooled in the furnace whilst for normalizing cooling is done in still air.

25 False.

After reading this section, the reader shall:

14 State the difference in composition, properties and use of copper, brass and bronze.
14.1 Identify brass and bronze as alloys of copper.
14.2 Recognize typical applications of copper, brass and bronze.
14.3 State the properties of copper.
14.4 Compare the properties of copper, brass and bronze.

Copper

Copper is a material which has been used for thousands of years.

It can be used either on its own as a commercially pure metal, or in combination with other elements as an alloy. The pure metal possesses desirable physical and mechanical properties:

(i) *High electrical conductivity*: copper is second to silver in conductivity, but the degree of purity of the copper is vital. Very small quantities of impurities can drastically reduce the conductivity – by as much as 25% in some instances.
(ii) *High thermal conductivity*: only silver conducts heat more readily.
(iii) *High ductility*.
(iv) *Good malleability*.
(v) *High corrosion resistance*.

Because of its high electrical conductivity copper has been used extensively as a conductor. In recent years, however, the cost of copper has risen so much that in some applications it has been replaced by aluminium, a typical example being overhead power cables. Whilst the conductivity of aluminium is lower than copper, its density is also much lower; and this is very important in applications such as overhead power cables.

Copper, due to its ductility and malleability, is easily worked. It may be processed by forging, drawing or rolling.

Production of copper may be by one of the following methods:

(i) *Electrolysis*: copper produced by electrolysis is most pure and is therefore used for conductors and other electrical applications. The purity achieved is a minimum of 99.9% copper.
(ii) In a *furnace*, where the impurities are lost as slag. Furnace-refined copper is used in food industries, in chemical engineering, for architectural cladding and for many other general engineering purposes.

In all the copper produced for commercial and industrial use the purity is above 99%. Copper in general has a low tensile strength. Where this property is important, then alloys must be used.

Brass

Brass is an alloy of copper and zinc. The zinc content can be as high as 45%. Sometimes other elements are added in small amounts, the more important additions being lead, tin, aluminium, manganese and iron.

When zinc is alloyed with copper, the percentage of zinc being less than 37%, then the atoms of the two materials bond together forming an alloy in such a way that the mixture appears as a new single material. These alloys are given the name 'Alpha' brasses (the symbol for them is the Greek letter α). Since the constituents bond in the way that they do, the α brasses have good toughness and ductility. The alloy with the highest ductility is the 70–30 brass (70% copper, 30% zinc). This is used in the manufacture of cartridges and shell-cases.

If more than 37% zinc is added, the structure of the alloy changes. Whilst most of it consists ofthe α phase as above, there are also parts filled by a hard brittle substance referred to as 'Beta' (the Greek letter β). β brasses have very poor ductility but greater strength than α brasses. Due to this low ductility the materials are difficult to work at room temperature.

The major differences between the two types of brasses are:

(i) α brasses can be hardened only by cold working, (this is similar to copper), but after hardening they can be annealed in the usual way.

(ii) α and β brasses can have their properties modified by heat treatment.

Brasses of each type can be further divided into more specific groups with a more precisely defined composition. Some of the more popular, together with their properties, are listed in Figure 183. The reader is not expected to remember the details in this table; it should only be used for reference.

Brasses with more than 45% zinc are very hard and brittle and are unsuitable for engineering structures. However, the increase in zinc reduces the melting temperature of the alloy, and this can be used as brazing spelter. The price of the alloy is cheaper, since zinc is less expensive than copper.

Bronze

Bronze, like brass, is an alloy of copper, but in this case the copper is combined with tin, not zinc. Tin is added in quantities up to 20%; usually there are also small amounts of phosphorus, zinc or lead. Since both copper and tin are expensive, the use of bronzes is limited. A large proportion of the alloys have less than 12% tin; when mixed in these quantities, the two materials combine to form α bronze. α bronzes are tough and ductile, and those containing 6% tin or less are suitable for cold working. Alloys containing between 6% and 12% can be cold-worked provided that they are correctly heat-treated in the initial stages of production.

Bronze alloys can be either wrought or cast. Wrought alloys contain up to 10% tin, whilst the cast alloys contain 10% or more tin.

The addition of phosphorus (in amounts up to 1%) increases the tensile strength and the corrosion resistance, and when used in a cast bearing alloy, it reduces the frictional resistance.

Types of brass

description	*type*	*composition*	*tensile strength* (MN/m²)	*elongation (%)*	*uses*
Copper-zinc brasses					
Gilding metal	α	90% copper 10% zinc	280 (annealed) 510 (hard)	55 4	Jewellery and decorative metal-work
Cartridge brass	α	70% copper 30% zinc	320 (annealed) 700 (hard)	70 5	Shell and cartridge cases. Deep drawing and presswork
Standard brass	α	65% copper	320 (annealed)	65	General-purpose alloy
		35% zinc	700 (hard)	4	Presswork, deep drawing extrusions
Basic brass	α	63% copper	340 (annealed)	55	General-purpose alloy
	(some β may be present)	37% zinc	725 (hard)	4	With limited application. A cheaper brass for presswork. Possible variation of properties
Yellow or Muntz metal	α and β	60% copper 40% zinc	370 (hot rolled)	40	Good corrosion resistance. Used for plates, rods and tubes, extrusions, pump components
Brasses with additional elements					
Free-cutting brass	58% copper 39% zinc 3% lead		450 (extruded rod)	30	Good machining qualities only suitable for limited cold-working
Naval brass	62% copper 37% zinc 1% tin		420 (extruded)	35	Forging and structural use. Improved corrosion resistance. Suitable for sand- and die-casting
High-tensile brass	58% copper up to 7% in total of aluminium, iron, tin and lead remainder – zinc		type A 470 min. B 540 min.	 20 15	Stampings and pressings. Wide use in marine engineering – propellers, rudders, castings, etc.

Figure 183 *Types of brass*

description	*composition*	*condition*	*tensile strength* (MN/m²)	*elongation (%)*	*uses*
Phosphor bronze	94.3% copper 5.5% tin 0.2% phosphorus	soft hard	350 700	65 15	Springs, instruments and steam-turbine blades
	89.5% copper 10% tin 0.5% phosphorus	cast	290	15	Bearings (small)
	81.5% copper 18% tin 0.5% phosphorus	cast	185	3	Heavy-duty bearings. Gear wheels. Worm gears
Gunmetal	88% copper 10% tin 2% zinc	cast	300	15	Good casting qualities. Pumps, valves, marine castings (good resistance to corrosion)
	85% copper 5% tin 5% zinc 5% lead	cast	225	13	As above but suitable where joints to withstand high pressures are necessary

Figure 184 *Types of bronze*

Like brass, the bronzes can be subdivided into more closely defined groups (Figure 184).

Sometimes lead is added, up to 20%, to improve the load bearing qualities of the bronze.

Another popular range of bronzes is the aluminium bronzes (Figure 185). The quantity of aluminium can be up to 14%, but it is very rare commercially to use an alloy with more than 10%. These materials can be either wrought or cast. They can be divided into hot-working and cold-working alloys. Alloys containing up to 5% aluminium are ductile and malleable, having a uniform structure, and they can readily be cold-worked.

The hot-working alloys contain up to 10% aluminium and when heated to approximately 800°C, they are very malleable. Castings are produced either by sand or die methods, but it is important that cooling is rapid to prevent brittleness.

Aluminium bronzes have good mechanical strength, resistance to creep, corrosion and wear resistance.

description	*composition*	*condition*	*tensile strength* (MN/m^2)	*elongaton* (%)	*uses*
Cold-working	95% copper 5% aluminium	soft hard	390 750	70 4	Imitation jewellery, sheet, strip, tubes (good resistance to corrosion)
Hot-working	7.5% aluminium, iron, manganese and nickel up to 2% total. Remainder – copper	hot-worked	425	45	Good qualities at high temperatures. Use in chemical engineering
	80% copper 10% aluminium 5% iron 5% nickel	forged	725	20	Marine propeller shafts. General marine applications. Can be heat-treated
	9.5% aluminium 2.5% iron, nickel and manganese up to 1% each. Remainder – copper	cast	525	30	Used extensively for sand and die casting. Chemical and marine engineering. Pumps, valves, casings, etc.

Figure 185 *Types of aluminium bronze*

Self-assessment questions

26 List any three applications of copper used in the home.

27 Which is the most desirable property of the purest copper obtainable?

28 List two physical properties and any two mechanical properties of copper.

29 What is the difference in the composition of copper and brass?

30 Explain why brass is not as good an electrical conductor as copper.

31 Select the material from list B which would be most suitable for each application in list A.

	A		*B*
(a)	Domestic water tank	1	Brass (α and β)
(b)	Marine rudder	2	Brass (high tensile)
(c)	Cartridge cases	3	Copper
(d)	Pump components	4	Brass (α)
		5	Brass (β)

32 From the list below select those properties which apply to copper.

(a) Brittleness
(b) Good electrical conductivity
(c) High tensile strength
(d) Ductility
(e) Low thermal conductivity
(f) Malleability

33 State the difference in composition of brass and bronze.

34 State two typical applications for
(i) Copper
(ii) Brass
(iii) Bronze

After reading this section, the reader shall:

15 Know the compositions, properties and recognize the uses of aluminium and magnesium alloys.
15.1 State the forms of supply, application and advantages of aluminium and magnesium alloys.
15.2 State the approximate compositions of some common aluminium and magnesium alloys.
15.3 List the four forms of aluminium alloys.
15.4 Identify the heat treatments necessary for aluminium alloys.
15.5 Outline applications of some typical aluminium and magnesium alloys.

Aluminium alloys

Aluminium is one of the light metals, having approximately one-third the density of steel. It has good corrosion resistance, high thermal conductivity and good electrical conductivity. Whilst its electrical conductivity is only half that of copper, weight for weight aluminium is a better proposition than copper. One advantage of aluminium is its high malleability which makes it possible to produce the very thin foil used in food packaging.

Solutions to self-assessment questions

26 Typical applications include: domestic water tanks, tubing for central heating, ornamental canopies above fires, jugs, vases.

27 Electrical conductivity.

28 Physical – electrical and thermal conductivity.
Mechanical – ductility and malleability, and corrosion resistance.

29 Copper is a commercially pure metal.
Brass is produced by the addition of zinc to copper in quantities up to 45%.

30 The addition of zinc to produce brass, together with other elements and impurities, drastically reduces the conductivity of the material.

31 (a) 3
(b) 2
(c) 4
(d) 1

32 (b), (d), (f)

In contrast to these good properties, aluminium has a major disadvantage: it lacks strength. Due to this, the material is of no use in constructional applications requiring this particular property. To overcome the problem aluminium is therefore alloyed with elements such as magnesium, manganese, copper and silicon. These are the main additives but others such as nickel, iron, zinc and titanium are sometimes added to produce specific properties.

Aluminium alloys can be produced in two forms – wrought alloys and cast alloys. Each of the two forms of alloy can be either a heat-treatable or non-heat-treatable type.

Non-heat-treatable wrought alloys

These alloys do not respond to strengthening by heat treatment; they can however be softened by annealing and then cold-worked (the amount carefully controlled), to produce the required strength and hardness. The alloys divide into three groups:

(i) 99% minimum pure aluminium with a total of 1% additions (the additions may be manganese, silicon, iron).
(ii) Aluminium with up to 1.5% manganese.
(iii) Aluminium with up to 7% magnesium.

When the materials are initially alloyed the elements combine as one material, giving high ductility. An increase in the amount of alloying elements or cold-working, which hardens and strengthens the alloy, causes the ductility to decrease. To prevent over-hardening and subsequent cracking the material may be annealed at a number of stages. Supply of the finished alloy is in various degrees of hardness ranging from soft, through ¼ hard, ½ hard, ¾ hard, and fully hard.

The addition of manganese strengthens the alloy whilst magnesium increases the resistance to corrosion.

Figure 186 gives the properties of principal aluminium alloys.

Heat-treatable wrought alloys

Most of the properties obtainable in these alloys are produced by heat treatment. The alloys divide into three groups:

(i) Alloys containing approximately 4% copper together with smaller amounts of silicon and magnesium.
(ii) Alloys containing no copper but with up to approximately 2% of silicon and magnesium.
(iii) Alloys containing various amounts of copper, silicon and magnesium – the copper is usually below 4%.

composition	*condition*	*tensile strength* (MN/m^2)	*elongation* (%)	*uses*
99% aluminium 0.1% max. manganese 0.5% max. silicon 0.7% max. iron	soft full hard	90 155	35 5	Decorative panelling, foil kitchen and hollow-ware, food and chemical industry, electrical conductors
98.8% aluminium 1.2% manganese	soft	110	33	Cooking vessels, general sheet metal work, scaffolding tubes, metal boxes
98% aluminium 2% magnesium	soft ½ hard	180 230	20 5	Ship and small boat construction, panelling, wire, deep drawn components
96.5% aluminium 3.5% magnesium	soft ¼ hard	220 275	18 8	Shipbuilding, deep drawing and pressing
95% aluminium 5% magnesium	soft ¼ hard	270 285	18 8	Shipbuilding, rivets, general marine applications requiring higher strength and corrosion resistance
Aluminium + 10% silicon 1.6% copper	chill-cast	250	3	Medium-strength general-purpose alloy, good for die-casting
Aluminium + 5% silicon 3% copper 0.5% manganese	sand-cast chill-cast	150 192	4 –	Sand-casting, die-casting. Inexpensive alloy. Good general-purpose alloy – low mechanical properties
Aluminium + 4.5% magnesium 0.5% manganese	sand-cast chill-cast	175 200	5 10	Sand and gravity die-casting, good marine applications. Suitable for medium stress, polishes well
Aluminium + 12% silicon	cast	115 to 205	5 15	Excellent foundry qualities. General and marine applications. Radiators, gearboxes, sumps

Figure 186 *Types of aluminium alloy*

Solutions to self-assessment questions

33 Both are alloys of copper. Brass is mainly a combination of copper and zinc, whereas bronze is an alloy of copper and tin. Each can contain other elements in smaller quantities.

34 Typical applications include:
Copper – electrical conductors, pipes, architectural cladding, boiler tubes.
Brass – jewellery, presswork and marine applications.
Bronze – bearings (light and heavy duty), pumps, valves, etc., coins.

composition	*condition*	*tensile strength* (MN/m^2)	*elongation* (%)	*uses*
92.5% aluminium 4% copper 1.52% magnesium 2% nickel	cast	280	18	(γ alloy) Pistons and cylinder heads for engines. Use on aircraft engine parts
91.5% aluminium 3% copper 5% silicon 0.5% manganese	cast	320	1	Suitable for all types of casting – general-duty alloy
87.6% aluminium 0.4% magnesium 0.50% manganese 11.5% silicon	cast	295	2.5	Good casting qualities, suitable for complex components. Good corrosion resistance but unsuitable for shock loading
90% aluminium 10% silicon	cast	320	18	Good applications in marine work because of good corrosion resistance. Good strength and ductility

Figure 186 continued *Types of aluminium alloy*

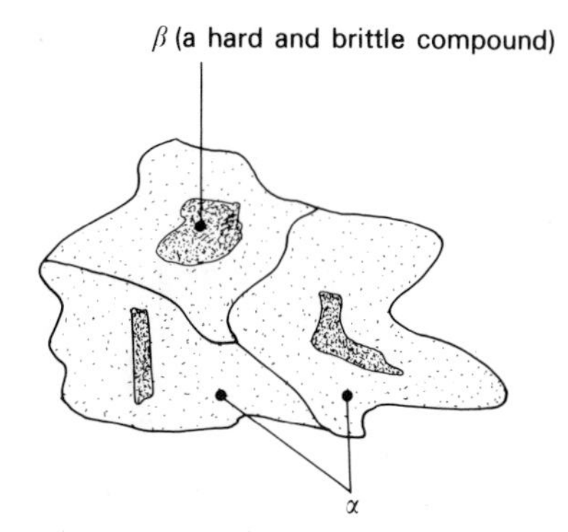

Figure 187 *Structure of a heat-treatable alloy*

Due to the possibilities of heat treatment these alloys tend to be stronger than the non-heat-treatable alloys. If the alloys are allowed to cool from the liquid stage very slowly under equilibrium conditions then a hard and brittle material is formed. This is due to the fact that the structure of the material consists of two different items. The major part is called α, in which the elements have combined as one. Inside this are large, hard, brittle components called β. Normally β is a compound and is formed by the precipitation of elements from the basic mixture. The structure resembles that shown in Figure 187.

Since the structure is brittle it is of very little practical use unless it has been subjected to heat treatment. This consists of heating up the alloy until all the constituents re-combine as one material again (*solution treating*), and then the material is rapidly quenched to room temperature retaining the uniform structure. Figure 188 shows the region into which the alloy needs to be heated before quenching.

After quenching, the alloy is very soft and in a similar state to an annealed steel. However, if left at room temperature for a few days the harder brittle component β forms once again, but in many small areas instead of a few large areas. This process is called natural *ageing* and the effect is to give strength and hardness to the alloy. The strengthening and hardening can also be achieved 'artificially' by heating the alloy, after quenching, at approximately 150–160 °C for several hours. This process is termed *precipitation* hardening. One great advantage of these alloys is the high strength:weight ratio achieved by the above treatments. Aluminium alloys are produced having higher strength:weight ratio than steel and find applications in, for example, the aircraft industry.

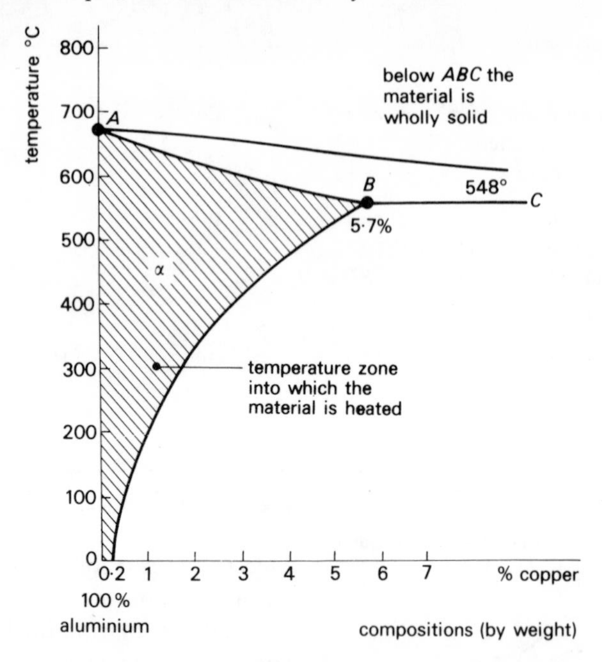

Figure 188 *Part of the aluminium-copper equilibrium diagram*

The range of alloys produced in this group is very extensive, giving a wide range of properties and finding many applications. Aluminium and copper (usually around 4%) alloys (plus other elements in smaller quantities) are termed Duralumin and can be used for structural purposes and stressed parts. With small variations in the addition of elements they can be used as highly stressed parts, such as engine parts e.g. connecting rods, for aircraft skin construction, etc. Alloys containing mainly zinc, copper and magnesium have the highest strengths, in the region of 600 MN/m^2. These parts are used for aircraft structures, military equipment, etc.

In general the other alloys find applications as glazing bars and window sections, structural members for road and rail vehicles, ladders and scaffold tubes, tubular furniture, rivets and general architectural work.

The material is obtainable in sheet, forgings, extrusions, tube and sections.

Also in this range are a number of special-duty alloys which have been specifically developed for service at temperatures up to 150 °C. One of special note is an alloy called Hiduminium – RR58 (composition – aluminium, 2.5% copper, 1.5% magnesium, 1.1% nickel, 1.1% iron and small additions of manganese and titanium) which is used extensively for aircraft skins.

Non-heat-treatable casting alloys

These alloys are used extensively for sand- and die-casting. They derive no benefit of strengthening by heat treatment. Alloys containing between 10% and 13% silicon are most important and are particularly useful for die-casting. High-casting fluidity and low shrinkage give added die-casting advantages. These materials also have high corrosion resistance.

Aluminium alloys containing magnesium and manganese are rigid and tough. These properties, combined with high corrosion resistance, make them very suitable for marine applications.

Heat-treatable cast alloys

As with wrought alloys the heat-treatable cast alloys are stronger than those which are non-heat-treatable. These alloys, many of which contain silicon, have other elements added which act as hardening agents. Many of the alloys are of the 4% copper type as described for wrought heat-treatable alloys.

One of the best-known alloys is the γ (gamma) alloy, developed during the 1914–18 world war. In addition to approximately 4% copper it also contains 2% nickel, 1.5% magnesium and other small additions. The composition gives it high strength and wear resistance at temperatures up to 150 °C and having good-casting qualities makes it suitable for pistons and cylinder heads for aircraft engines. It can also be supplied in wrought form.

Magnesium alloys

Magnesium, like aluminium, is very light; in fact it is the lightest metal, having a relative density of only 1.7 which is approximately ⅔ that of aluminium. A major problem with magnesium is its manufacture, since it is highly reactive when in contact with oxygen. Due to its reactive character it has been used in :ares and in incendiary bombs (particularly during the 1939–45 war). Magnesium also suffers from the same defect as aluminium: it is weak and unsuitable for general use, so alloys are formed. The structure of magnesium is such that the materials are difficult to work when cold but can be worked reasonably easily whilst hot. After alloying, the main advantage which is obtained is a high strength:weight ratio. Having this property its most important applications are in the aircraft industry or other areas where weight is a limiting factor.

The elements most commonly added are aluminium and zinc but, also, small quantities of manganese, silver, zirconium or other rare earth metals may be included. Good properties in the alloys are usually achieved by the same heat treatments as for aluminium. Most of the elements added help to strengthen the alloys but manganese improves the corrosion resistance, since pure magnesium is not particularly corrosion resistant.

Magnesium alloys (Figure 189) can be supplied in a number of forms: sand- or die-cast, forged, extruded, rolled in sheet or plate. It can be machined at high speeds and also argon-arc welded.

composition	*conditon*	*tensile strength* (MN/m^2)
91% magnesium 8% aluminium 0.3% manganese 0.7% zinc	cast and hardened	200
94.8% magnesium 4.5% zinc 0.7% zirconium	cast and hardened	260
93.9% magnesium 4.2% zinc 0.7% zirconium 1.2% rare-earth metals	cast and hardened	215
98.5% magnesium 1.5% manganese	rolled extruded	200 235
92.7% magnesium 6% aluminium 0.3% manganese 1% zinc	forged extruded	280 215
96.3% magnesium 3% zinc 0.7% zirconium	rolled extruded	260 305

Figure 189 *Types of magnesium alloy*

Typical applications of magnesium alloys are for lightweight road wheels (especially for racing cars), crankcases and gearboxes for aircraft and cars, portable tools, ladders, airframe components, camera cases, business machines and lawn mowers, etc. By using magnesium alloys in moving parts, power can be saved due to the lightness of the components.

Self-assessment questions

35 Aluminium alloys are usually supplied in four forms. What are the forms?

36 When aluminium is alloyed, which property can be achieved which makes the alloy a better proposition than steel?

37 The alloy is heated until all the constituents re-combine as one (α) and is then quenched rapidly to room temperature. What is the name of this treatment?

38 What condition is the material of self-assessment question 37 in after the treatment?

39 Give the names of two hardening processes used after solution treatment.

40 For the compositions below, state the form of supply of the alloy:
(i) 98% aluminium, 2% manganese
(ii) Aluminium, 4% copper, small amounts of silicon and magnesium
(iii) Aluminium, 10% silicon, 1.6% copper
(iv) 90% aluminium, 10% silicon

41 By which process do magnesium alloys achieve their maximum strength?

42 Give four elements commonly alloyed with magnesium.

43 Listed below are a number of elements combined to form alloys. State the approximate composition of each.
Alloy (a) – magnesium and manganese
Alloy (b) – magnesium, aluminium, manganese and zinc (give figures for both a cast and a forged material)
Alloy (c) – magnesium, zinc and zirconium (rolled or extruded material)

44 For which of the following applications are magnesium alloys suitable?
(a) Portable tools
(b) Bridge construction
(c) Machine beds
(d) Aircraft parts
(e) Ladders
(f) Railway tracks

Additional assessment questions – aluminium and magnesium and their alloys

45 All aluminium alloys can be heat-treated to produce their required strengths. TRUE/FALSE

46 Select from the list below the possible forms of supply of magnesium alloys.
(a) Sand casting
(b) Forging
(c) Rolled sections
(d) Extrusions
(e) Die castings

47 List six suitable applications for aluminium alloys.

48 Which is the lighter metal – aluminium or magnesium?

49 What are the disadvantages of using pure aluminium or magnesium structurally?

50 Magnesium alloys are unsuitable for cold-working.
TRUE/FALSE

After reading this section, the reader shall:

16 Know the compositions and properties of self-soldering alloys.
16.1 State the differences in composition, properties and uses of tin–lead alloys.
16.2 Identify the composition of tinman's and plumber's solders.
16.3 Recognize typical uses of solders.

Alloys of tin and lead are used extensively as solders for joints between other metals. The proportion of tin varies between 12%

Solutions to self-assessment questions

35 Wrought alloys – heat-treatable
Wrought alloys – non-heat-treatable
Cast alloys – heat-treatable
Cast alloys – non-heat-treatable

36 It achieves a higher strength:weight ratio and is therefore highly advantageous for aircraft parts.

37 Solution treatment.

38 A soft and ductile state – similar to an annealed steel.

39 Ageing and precipitation hardening.

40 (i) Wrought alloy – non-heat-treatable
(ii) Wrought alloy – heat-treatable
(iii) Cast alloy – non-heat-treatable
(iv) Cast alloy – heat-treatable

41 By heat treatment similar to aluminium alloys – ageing or precipitation hardening.

42 Aluminium, manganese, zinc, zirconium.

43 Alloy (a) – 98.5%, 1.5%
Alloy (b) – 91%, 8%, 0.3%, 0.7% (cast)
90.7%, 6%, 0.3%, 1% (forged)
Alloy (c) – 96.3%, 3%, 0.7%

44 (a), (d), (e)

Solutions to additional assessment questions

45 False

46 Magnesium alloys can be obtained in all forms listed.

and 62%, the remainder being lead. Sometimes silver and/or antimony may be added. Antimony gives strength to the solder. Since tin is far more expensive than lead the tin content is sometimes reduced.

One of the main reasons for the suitability of these alloys as solders is that they have a low melting point. Typical solders are listed in Figure 190.

composition of solder	*temperature range for solidification*	*names and general application*
62% tin 38% lead	183 °C	Tinman's solder (often called 'fine solder')
50% tin 50% lead	220 – 183 °C	Tinman's solder (coarse). Quantity of tin reduced for cost
33% tin 67% lead	260 – 183 °C	Plumber's solder
31% tin 67% lead 2% antimony	235 – 188 °C	Plumber's solder for wiping joints
18% tin 80.8% lead 0.8% antimony 0.4% silver	270 – 182 °C	Substitute for plumber's solder
43.5% tin 55% lead 1.5% antimony	220 – 188 °C	General-purpose solder
30% tin 69.5% lead 0.65% silver	250 – 182 °C	Substitute general purposes
40% tin 60% lead	236 – 182 °C	Tin-can solder
12% tin 80% lead 8% antimony	250 – 243 °C	For soldering steel and iron

Figure 190 *Types of solder*

The solders listed divide into two distinct areas, those used by a tinsmith or those used by a plumber. 'Fine' solder used by the tinsmith solidifies instantly at 183 °C. It therefore passes quickly from the liquid stage to the solid stage, reducing the possibility of disturbing and breaking joints which sometimes occurs during a pasty stage. The plumber's solders however have a pasty stage and solidify over a range of temperature. This is a definite advantage for the plumber since he is able to 'wipe' joints.

Self-assessment questions

51 What is the composition of 'fine' tinman's solder?

52 Give two other elements sometimes added to lead–tin alloys.

53 A solder has the following composition – 43.5% tin, 55% lead, 1.5% antimony. What is a typical application for this alloy?

54 Give the composition of solders having the following uses or applications:
(a) Plumber's solder
(b) For soldering iron or steel
(c) For soldering tin cans

After reading this section, the reader shall:

17 Be aware of those materials which have good bearing qualities.
17.1 Identify the most important bearing materials as phosphor bronzes, cast iron, PTFE, nylon, graphite and white metals.
17.2 State the individual properties of the materials listed above.
17.3 Identify some applications of the materials in **17.1**.

Many people are familiar with ball and roller bearings, which enable the free movement of one part relative to another. These bearings can be costly and all assemblies do not require the same ease of movement that they produce. An alternative is to use a solid material as the bearing, ensuring that the material used has suitable bearing qualities. Phosphor bronze, cast iron, PTFE, nylon, graphite and white metals are used in situations which call for their particular properties.

An advantage of some of these materials is that they are self-lubricating, and may therefore be used in applications where normal lubricating procedures are not possible.

The more important properties a bearing metal must possess are:
(a) It is hard; this prevents the bearing becoming mis-shaped due to loading.
(b) It is wear-resistant; this enables the running tolerances to be maintained.

Solutions to additional assessment questions

47 Aircraft parts – engine and frame, marine applications, rivets, radiators, gearboxes, sumps, scaffold tubes, cooking utensils, decorative panelling, etc.

48 Magnesium

49 They are very weak.

50 True

(c) It has a low coefficient of friction to allow free running.
(d) It is tough enough to withstand any shock loadings which may be applied to the bearings.
(e) It has sufficient ductility to allow for initial 'bedding-in' or 'running-in' of the bearing.

No pure metal has all the properties necessary to allow it to be used as a bearing. To achieve all or the majority of the properties it is necessary to use an alloy, a composite material or a polymer. The alloys have elements added so that the resulting material has hard particles (which are normally intermetallic compounds) set in a softer material. These combinations of materials meet most of the bearing requirements. The composites are much more complex, and are beyond the scope of this text. Polymers are now being used extensively, particularly where loads are light and where oil lubrication may be undesirable or difficult to use. Polymers are also relatively inexpensive.

Typical bearing materials and their application:

Phosphor bronze, an alloy of copper and tin with a small addition of phosphorus. The amount of phosphorus usually varies between 0.1% and 1.0%; it increases both the tensile strength and the corrosion resistance, whilst reducing friction. The higher amounts of phosphorus give the lower friction and are therefore most suitable for use as bearings, particularly in heavy-load-bearing conditions. Most of the tin combines with the copper creating hard particules; the remainder forms a softer but tough matrix throughout which the harder particles are randomly distributed.

Cast iron is used as a bearing material when the carbon in the cast iron is in the form of graphite, i.e. grey cast iron. The graphite, usually in flakes, makes the iron softer, easy to machine and has a low coefficient of friction. Cast iron can be used for bearings provided that speeds and loads are moderate or low. It is also essential to provide lubrication. The material is resistant to wear, cheap and easily obtained.

PTFE (polytetrafluoroethylene) has a very low coefficient of friction and feels greasy to the touch. The material is clean and quiet in use, and is highly resistant to corrosion. It is one of the more expensive polymers, and therefore only used when its special qualities are needed. The material is capable of operating at temperatures up to 300 °C. Sometimes other materials, graphite being an example, are added to improve its properties.

Nylon, one of the most commonly known polymers, has a low coefficient of friction. The material is strong, hard wearing, tough and suitable for working at temperatures up to approximately 140 °C. It is relatively easy to produce by injection moulding or by extrusion. It is, however, one of the more expensive polymers. Due to its properties and ease of production it is now widely used for bearings on cycles, domestic appliances, etc.

Graphite is one of the two basic forms of carbon, the other being diamond. The structure of the material is such that the atoms slide easily over each other, giving it properties suitable for bearing applications. Since graphite is weak it is not used in its pure state as a bearing. The common procedure is to use graphite within the structure of other materials. This can be done by

(a) Allowing the graphite to form on solidification e.g. in cast iron, or
(b) Using it as filling to vary the properties, examples being with plastics, etc., or other metals

White metals are alloys primarily of tin, lead and antimony. They can be either tin-based or lead-based, and both types usually contain approximately 10% of antimony. The resulting structure is of hard cuboids within a much softer and tougher matrix. It is thus a combination of the desirable bearing properties discussed earlier. The tin-based alloys are used in demanding situations such as the main bearings in car and aero engines. Whilst the addition of lead cheapens the material, the resulting materials are suitable only for light duties, involving low pressures and speeds.

In addition to the bearing materials discussed, a common type used is a copper-based alloy produced by sintering. This is a method of producing components by compressing powders and then heating them to pre-determined temperatures. Sintered materials are semi-porous. The cavities in a component are filled with oil which makes the bearing self-lubricating. Bearings made from sintered materials are used in light-duty applications only, particularly where little or no attention is expected during the life of the equipment. Sometimes carbon (in the form of graphite), instead of oil, is used to fill the cavities.

Solutions to self-assessment questions

51 62% tin, 38% lead.

52 Antimony and silver.

53 It is a general-purpose solder.

54 (a) 33% tin, 67% lead
(b) 12% tin, 80% lead, 8% antimony
(c) 40% tin, 60% lead

Self-assessment questions

55 From the list below select four properties which a bearing metal must possess.
(a) High coefficient of friction
(b) Hardness
(c) Brittleness
(d) Some ductility
(e) Wear resistance
(f) Toughness

56 A planing machine is to be made from some material such that the table slides easily over the bed. If the table and bed are to be similar materials, suggest a suitable material from those discussed in the text.

57 Pure metals possess all the necessary properties for a bearing.
TRUE/FALSE

58 When two metals are combined to produce a bearing material they usually have a common type of structure. What is the structure like?

59 Which of the following materials are NOT commonly used as bearing materials?
(a) High-carbon steel
(b) Cast iron
(c) Magnesium
(d) Phosphor bronze
(e) An alloy of lead, tin and antimony
(f) Aluminium

After reading this section, the reader shall:

18 Understand the difference in thermoplastic and thermosetting plastic materials.
18.1 State the difference between thermoplastic and thermosetting plastics.
18.2 Identify thermoplastic and thermosetting materials.
18.3 State the engineering uses and properties of typical thermoplastic and thermosetting plastics.

Plastics are now finding many applications in all types of industry. They can be divided into two main groups:
(i) Thermoplastics
(ii) Thermosetts

Thermoplastics

These materials can be softened by heating but become hard again

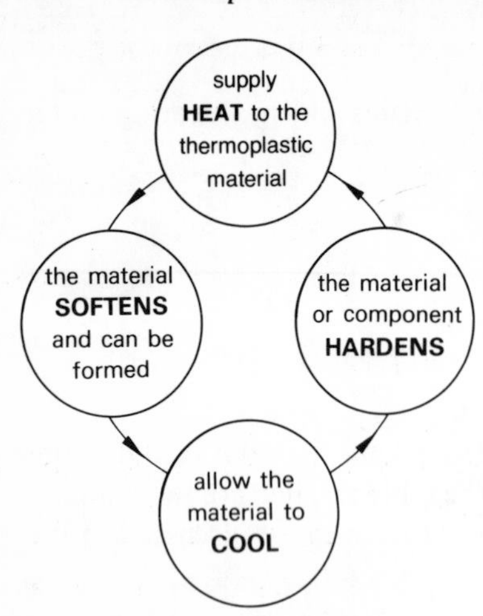

Figure 191 *Behaviour of thermoplastics*

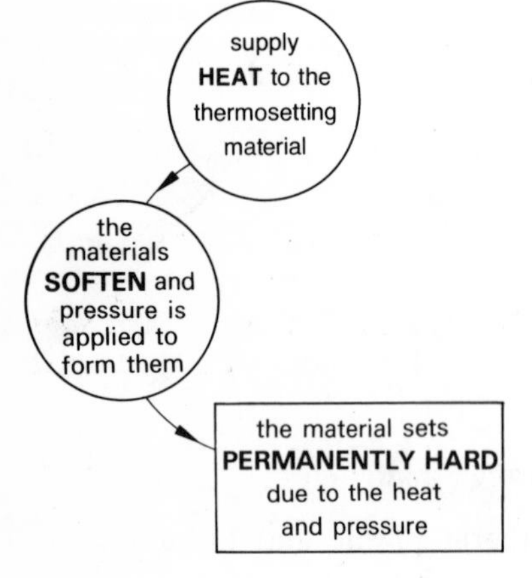

Figure 192 *Thermosetting plastics*

when cooled. The process can be carried out repeatedly. However, too-frequent processing causes the quality of the material to deteriorate. Since no chemical change takes place, materials can be recycled. This allows savings to be made in raw materials.

The thermoplastic materials are heated, and when soft, formed into the shape of the final component by the use of dies or moulds, and then allowed to cool. On cooling the thermoplastic retains the shape into which it has been formed, being stiffer and more rigid than in the forming stage.

The raw materials are usually in powder or granular form.

One disadvantage of thermoplastics is that operating temperatures must be carefully controlled, or the component may soften and lose shape. However, there are many thermoplastic materials, so they can be used over a wide range of operating temperatures for different uses.

The behaviour of thermoplastics can be represented by the diagram in Figure 191. The circuit in the diagram can be followed round continually.

Thermosetts

Unlike thermoplastic materials the thermosetts cannot be softened and recycled. To produce a component in thermosetting material, powder or granules are used which have a structure similar to a thermoplastic material. The material is placed in dies and heated. During the heating, pressure is applied to the powder or granules. The effect of heat is to soften the material initially; the pressure forms the component and the heat eventually 'cures' the material with the result that it becomes permanently hard. During the 'curing' a chemical reaction takes place making the material hard and rigid. The structure is made permanently rigid by the chains of atoms and molecules cross-linking with each other during the reaction to form a complex network. The process can be represented as in Figure 192.

Solutions to self-assessment questions

55 (b), (d), (e) and (f)

56 The best material for machine beds which has bearing qualities is cast iron.

57 False. To obtain the very wide range of properties, materials need to be combined together.

58 The main structure usually consists of a soft but tougher material as the body. This contains hard particles of the second material.

59 (a), (c) and (f)

Some of the more common and widely used thermoplastics are:

1 Polyethylene
2 Polypropylene
3 PVC (polyvinyl chloride)
4 Polystyrene
5 PTFE (polytetrafluoroethylene)
6 Nylon
7 Acrylics (polymethyl methacrylate)
8 ABS (copolymers of acrylonitrile, butadiene and styrene)

Thermoplastics are produced in a variety of different forms. For instance density can be changed or the flow properties varied. This gives materials the properties needed to meet the ever-changing requirements of industry and society. Methods are available which can, for instance, strengthen, toughen, or cheapen the materials so that a plastic can almost be 'tailor-made' to meet a specification.

Thermoplastic materials

1 *Polyethylene*, more commonly called polythene. The material may be produced with either a high or a low density. The main difference is that the high-density material is more rigid, has more strength and withstands higher temperatures.

Polyethylene has a high resistance to corrosion caused by solvents and chemicals. It is unaffected by food. The material is tough and flexible, and since it has a low softening temperature can be readily formed into desired shapes by a number of processes. It is cheap to produce.

Some common uses of polyethylene are packaging (in the form of thin films or sheet for sweets, food, frozen food and other products), industrial and household containers, toys, cold water pipes and electrical insulation.

2 *Polypropylene* is a similar material to high-density polyethylene but with more rigidity, strength and heat resistance.

Polypropylene has better resistance to solvents and chemicals than polyethylene. One distinct advantage is that containers made in this material can be sterilized by boiling without detriment to the material. It also has the ability to withstand repeated bending without cracking, and hence is sometimes used for hinges and accelerator pedals in cars.

Polypropylene is widely used in:

(i) Films
(ii) Packaging
(iii) Electrical insulation
(iv) Automobile parts (trims, wheel arches, pedals, etc.)

(v) Cases and containers (crates for milk and beer bottles, battery cases, etc.)
(vi) General household and industrial containers
(vii) Safety helmets

3 PVC (*polyvinylchloride*) can be produced in either a rigid form or a flexible form. The rigid form is much more difficult to work and produce. It is hard when solid and possesses both strength and toughness. To increase the range of the properties of the material and ease production a plasticizer is added. This results in flexible forms being produced, the final properties being dependent upon the amount of plasticizer added. The flexible materials range from a tough, pliable material to a very soft, flexible and rubbery type of material. These additional properties are accompanied by a loss in strength. The majority of products produced in PVC are formed using plasticized material.

In addition, vinyl chloride can be combined with vinyl acetate and other compounds to produce a further range of similar materials to PVC. The variety of combinations which can be produced is extensive, giving tremendous versatility of properties. The materials have exceptional resistance to acids, solvents, etc., and have excellent insulation properties.

Some common uses are moulded helmets (from rigid material), bottles, handbags, protective gloves and clothing, garden hose, floor tiles, luggage, electrical insulation, records, tank linings, interiors of refrigerators, coverings for upholstery.

When the appropriate plasticizers are added the material can be sprayed, or used for coating metal as a protective by dipping.

4 *Polystyrene* is a very widely used material and now one of the most important thermoplastics. It can be produced in clear form with a hard, glossy surface. The material is strong, flexible and light in weight, but it has a low impact strength. It can be easily moulded into shape.

Polystyrene has very high electrical resistance and good resistance to foods and household acids. An advantage of the material is that it can be combined with other compounds to produce a large range of other materials with varied properties. The impact strength can be improved, and also the resistance to chemicals and oils.

Some of these plastics can be produced in the form of foams by releasing air or gas bubbles within the plastic during solidification. Foamed polystyrene called 'expanded polystyrene' can be produced containing up to 97% air. Because it is light, polystyrene is made into ceiling tiles; these have excellent thermal qualities.

Some typical uses of the range of polystyrene are for toys, refrigerator parts, packaging for foodstuffs, thin-walled containers, packaging for fragile parts (expanded polystyrene), trays, pens, hulls for some sailing boats, thermal and sound insulation in houses and factories (expanded).

5 PTFE (*polytetrafluoroethylene*), due to its structure, resists attack by most corrosive chemicals and solvents. It is unaffected by hot concentrated sulphuric acid and other acids which are capable of dissolving some metals. PTFE can be used at temperatures ranging from as low as −200 °C up to 300 °C with little effect on its properties. Most materials do not adhere to PTFE, and this property makes it very suitable for non-stick pans and similar applications.

The material is tough, is flexible, is an excellent electrical insulator, and has a very low coefficient of friction; however, its tensile strength is low.

Typical applications are small light-duty bearings, non-stick surfaces, gaskets and sealing rings, valve seats, piston rings, electrical insulation, protective coatings, pipes and joints, pump parts.

PTFE is one of the most costly thermoplastics produced, and cannot be manufactured by the conventional techniques used for thermoplastics. It is produced by heat and compaction in a fashion similar to metals in powder metallurgy. This is because the material does not soften and flow when heated as is normal with thermoplastics.

6 *Nylon* is the name given to a group of materials termed 'polyamides'. Its range of properties makes it attractive to the design engineer, and it is widely used. Nylon is very strong, tough, durable, possesses good resistance to abrasion, is light in weight, and has a low coefficient of friction. It is flexible and has high impact strength and also good insulating properties. Nylon has the important property that it can be used without lubrication.

Some polyamides are filled with fibres (usually glass). This improves the mechanical strength. Graphite may be added to reduce the frictional resistance. Nylon resists attack from most chemicals and solvents, but is damaged by attack from some mineral acids. It has a reasonably high softening temperature (approximately 200 °C), and is therefore safely used at operating temperatures up to approximately 150 °C.

Some nylons tend to absorb water with a resulting reduction in strength and impact properties. This absorption of water may cause changes in the dimensions of a component.

Uses of nylon include small gear assemblies, bearings, fibre and yarn for use in hosiery, etc., combs, bristles for brushes, ropes, fishing lines, tennis racquets and strings, casing and machine parts for domestic appliances, couplings, helmets, tubing, insulating clips, conveyor belt rollers, pulleys, transistor cases, carburettor floats.

7 *Acrylics*. One of the most common polymers in this group is polymethyl methacrylate, which is a clear glass-like material. It is commonly called *perspex*. The material is hard and rigid with high impact strength and good tensile strength. However, one disadvantage is that the surface is easily scratched. It is resistant to most household chemicals but is attacked by acetone, petrol and turpentine. As with many polymers its electrical resistance is high.

Perspex transmits more than 90% of daylight and is tougher and lighter than glass. It is also easily moulded into shape having a low softening temperature (approximately 110 °C).

Perspex is used for aircraft glazing, roofing panels (windows), telephones, baths, sinks, display signs, building panels, lenses, drawing instruments, clock faces, handles, knobs, rear light lenses and reflectors.

Other materials in the group such as polyacronitile are produced as fibres, and used extensively in clothing manufacture. (Some of the well-known trade names of these materials are 'Acrilan', 'Orlon' and 'Courtelle'.)

8 *ABS* plastics are co-polymers (combination of a number of polymers) of acrylonitrile, butadiene and styrene. These materials have exceptional resistance to impact with a good tensile strength. They are resistant to most acids, to many alkalis and to some solvents. Having a low softening temperature they can only be used up to a maximum operating temperature of approximately 80 °C. They are easily processed by a number of the methods commonly used for thermoplastics.

ABS is used for pipes, protective helmets, tool handles, toys, wheels, refrigerator parts, pump components, luggage, radio cases and battery cases. It is also used widely in the automobile industry for parts of bodywork (some of which may be chromium plated).

Thermosetting materials

The most common thermosetts are listed below:

1 Phenolics (phenolformaldehyde (bakelite))
2 Urea formaldehyde
3 Melamine formaldehyde
4 Polyurethane
5 Epoxy resins
6 Polyester resins

The range and variety of thermosetts is far less extensive than for thermoplastics.

1 *Phenolics*. The commonest material is phenolformaldehyde, better known as bakelite. It was the first thermosetting material produced commercially, being discovered in 1909. The material is hard, strong and rigid and able to operate at temperatures up to 200 °C. One of its advantages is that it has high electrical resistance, and due to this, it has developed alongside the electrical industries being used for electrical fittings such as plugs, etc.

The material has high chemical resistance and dimensional stability.

Bakelite is used for many purposes including electrical fittings, knobs, buttons, switchgear, saucepan handles, motor car parts, cabinets and mouldings for radios, ignition systems and parts for domestic appliances.

2 *Urea formaldehyde* materials are similar to phenolformaldehyde. They are naturally colourless, but pigments are usually added and provide a variety of colours. The surface of the materials is very hard and scratch-resistant, but the maximum operative temperature is only approximately 80 °C. These materials may be strengthened and toughened by the addition of other substances. They have high electrical resistance and are resistant to solvents and chemicals (including petrol and oil). A disadvantage is that water is usually absorbed when it comes into contact with the material. This results in a loss in dimensional stability.

These materials are used in electrical plugs and sockets, buttons, bottle tops, radio cabinets, kitchen equipment, surface coatings, as a bond for foundry sand, toilet seats.

3 *Melamine formaldehyde* materials have some properties similar to urea formaldehyde, but have a higher service strength and greater hardness. (They are perhaps the hardest of all plastics.) They are also much more resistant to water absorption than urea formaldehyde. The improvements in the properties result in the materials being suitable for the manufacture of cups, saucers, plates and kitchen equipment, particularly where heat resistance is required. These materials are also used for decorative panels (formica, etc.), electrical fittings (for which it has superseded 1 and 2), aircraft distributor and ignition parts, trays, light fixtures, radio cabinets and handles.

4 *Polyurethanes*. One type is used in the manufacture of bottles, films and filaments. However, many types are used as foams. The foams range from a hard and rigid material to a soft and flexible material.

The rigid foams are generally used for heat insulation and for strengthening hollow structures by producing a sandwich construction. The soft and flexible foams are used for upholstery (particularly in cars) and for items such as artificial sponges.

Polyurethanes are also used in adhesives (usually glass to metal), in bases for paints, and in electronic equipment. They are also used in the manufacture of knobs and handles.

5 *Epoxy resins* have a complex structure which may be arranged during manufacture to give properties suitable for many applications. The resins are usually combined with a hardener; one form of hardener is cured and becomes hard at room temperature; the other form requires heat for curing over a longer period of time.

The resulting properties of the materials are:

(a) High mechanical strength and toughness
(b) High dielectric strength and resistance
(c) Excellent adhesion to many other materials, including metal, glass and plastics
(d) High resistance to chemicals and moisture
(e) Negligible shrinkage
(f) Excellent wear resistance

In addition, these resins are usually combined with fibres (such as glass or carbon) to produce hard and strong structural materials.

Epoxy resins are expensive, but are widely used in circumstances that demand their particular properties. They are used in the casting of certain components and in the production of laminates where the strength to weight ratio is important. They are used in bonding of materials and in surface coatings. The trade name of a common impact adhesive which is an epoxy resin glue is 'Araldite'.

A widely used application is to insulate and encapsulate electrical and electronic components and systems. In the encapsulating process the component is sealed within the material.

Flooring made from epoxy aggregates is popular due to the wear resistance and chemical resistance of the material.

6 *Polyester resins* cover a large range of materials; the most commonly used types are laminating and casting resins. Many cure at room temperature but heat can be used to accelerate the process. As with epoxy resins, these materials are combined with fibres (glass usually), to produce a range of materials suitable for many applications. The final properties depend upon:

(i) The basic polymer being used, and
(ii) The quantities and methods of combining the fibres

The glass-fibre polyester resins are very strong, durable and have good impact resistance. They are resistant to solvents, dilute acids and alkalis, but may be attacked by strong acids. The materials are good electrical insulators.

Polyester resins are used in the manufacture of hulls of small boats, wheelbarrows, chairs, car bodies, fishing rods, adhesives and surface coatings. They are also used sometimes in the insulation of electrical equipment.

Self-assessment questions

60 Name one property which is common to most plastics.

61 Which of the following thermoplastics has the highest operating temperature?
(a) PVC
(b) Nylon
(c) PTFE
(d) Polystyrene

62 List three thermoplastics which may be used for household containers.

63 State the main difference between a thermoplastic material and a thermosetting material.

64 State which of the following materials are thermosetts and which are thermoplastics.
(a) Epoxy resin
(b) Nylon
(c) Urea formaldehyde
(d) Polypropylene
(e) Polyurethane
(f) Polyethylene
(g) Polystyrene

65 Which thermosetting material is used for cups, saucers, plates etc?

66 Suggest thermosetting materials suitable for the following applications. Note that in some instances more than one material is suitable.
(a) For bonding metal to glass
(b) Electrical plugs and sockets
(c) Reinforcement for aircraft wings
(d) Buttons
(e) Boat hulls

67 State four of the outstanding properties of epoxy resins.

68 Thermoplastic materials are generally stronger and more rigid than thermosetting plastics.

TRUE/FALSE

69 State four plastics commonly used in the home, together with their application.

Solutions to self-assessment questions

60 High electrical resistance. This makes them suitable for use as insulating material.

61 (c) PTFE

62 The list may include polypropylene, PVC, polystyrene, polythene.

63 A thermoplastic can be repeatedly heated, softened and re-hardened again without detriment to its properties, whilst a thermosett becomes *permanently* set during its initial production.

64 Thermosetts (a), (c), (e)
Thermoplastics (b), (d), (f), (g)

65 Melamine formaldehyde

66 Typical solutions are:
(a) Polyurethane or epoxy resin
(b) Phenol formaldehyde, urea formaldehyde
(c) Polyurethane foam
(d) Phenol formaldehyde, urea formaldehyde
(e) Epoxy and polyester resins

67 Any four from the following:
(a) High strength and toughness
(b) High dielectric strength and resistance
(c) Excellent adhesion
(d) Resistance to chemicals and moisture
(e) Shrinkage is negligible
(f) Excellent wear resistance

68 False. Thermosetts are stronger and more rigid than thermoplastic materials.

69 Some typical solutions are:
Bakelite – for electrical plugs
Polyurethane – knobs, handles, paint
Nylon – domestic appliances
Acrylics – clothing
Other examples are polyethylene (in food packaging), melamine formaldehyde (formica, etc.)

Topic area: Protection of metals

After reading the following material, the reader shall:

19 Know the reason for and the method of protecting metals from atmospheric attack.

19.1 State the need for protection of metal from atmospheric corrosion.

19.2 Identify methods of painting metal surfaces and the hazards inherent in spraying and stoving.

19.3 Identify the principles involved in a range of protective treatments.

19.4 Select appropriate protective treatment (e.g. anodizing or plating) for specific metals.

Almost all metals, when in contact with the atmosphere, eventually show signs of corrosion. The rate of corrosion varies with different metals. Some metals have high resistance to corrosion, but materials such as iron and steel corrode quickly and need to be protected.

The process of corrosion is a chemical reaction, called chemical corrosion. Some metals combine with the oxygen in the atmosphere and an oxide film forms on the surface. This film may 'flake off' or it may be porous; in each case the metal underneath the oxide film continues to corrode. Alternatively the oxide may be protective preventing further corrosion, as with aluminium.

Materials such as iron and steel need the presence of oxygen and moisture before corrosion can commence. In the presence of dry air there is no reaction. When the moisture contains salt then the corrosion is more rapid. A common example of steel corrosion is the rusting of motor cars. If the material is allowed to rust freely holes soon appear in the car body.

Another form of corrosion is called electro-chemical corrosion. This is again a chemical reaction but of a more complex type. Electrolytic action takes place when dissimilar metals are in close proximity to each other and are also covered or in continuous contact with some electrolyte.

The corrosion reaction is due to the electric currents occurring as a result of the potential difference which is set up between the metals in contact with electrolyte. The amount of current or the rate at which corrosion takes place depends upon the electrode potential of the metals. Materials are listed and arranged in accordance with their electrode potential above and below that of a standard

carbon	↑ positive potential difference
gold	
platinum	
silver	
mercury	
copper	
hydrogen	
lead	↓ negative potential difference
tin	
nickel	
cadmium	
iron	
chromium	
zinc	
aluminium	
manganese	
sodium	
potassium	
lithium	

Figure 193 *The electrochemical series*

electrode, hydrogen. The ones above the standard are regarded as having positive potential differences and the ones below the standard as negative potential differences. The list is called the electrochemical series and some comparisons are shown in Figure 193.

When two materials in the list are used together, the further apart they are in the list, the faster the rate of corrosion is likely to be. This is therefore an important consideration in design.

It is necessary to coat the surfaces of the materials impeding the corrosion process. Some of the methods are explained below.

Protection of metals

Painting

One of the most common methods of protecting metal against corrosion is to paint the surfaces. Most paints are not completely impermeable to water and hence do not prevent corrosion. If the paint surface is damaged, corrosion may occur rapidly (e.g. motor cars). To give adequate protection a priming paint is first applied; this is followed by an undercoat, and finally one or several finishing coats are added. The priming coat is usually a special type of paint which coats the surface, acting as a key for the undercoat. Sometimes special treatments may be necessary before the priming coats, which dissolve any grease or initial corrosion, and also prepare the surface for painting.

Paint can be applied to materials by:

(a) Brushing
(b) Spraying
(c) Electro-static spraying
(d) Dipping

For large quantities it is economically advantageous to spray or dip the components. This enables the process to become semi-mechanized, keeping cost down.

The drying time for paints varies, but many new paints are 'touch-dry' in minutes. Sometimes components or products are painted and then placed in a hot oven to dry. This process is called stove-enamelling. Products coated by this method have a very hard, wear-resistant finish. The most common application is on domestic appliances such as cookers, washing machines, refrigerators and sinks.

The use of paint creates workshop hazards. The most common and important are:

(i) *Fire hazard*. The constituents of paints can be highly inflammable. This means that all areas in which paint is used or stored must be 'no smoking' areas. It is essential to have fire extinguishers readily available, and where possible install sprinkler systems.

(ii) *Fire hazard when stoving*. Stoving of the components causes the thinners to evaporate very quickly. This results in inflammable fumes becoming quite extensive within the oven. Control and extraction is vitally important to prevent fire and possible explosions.

(iii) *Fume hazard*. Many paints give off fumes which may be dangerous. Since most painting is done in special booths, they must be very well ventilated, providing a satisfactory and safe space in which to work. Some of the heavier particles in the paint become airborne and are difficult to remove. To prevent breathing these particles it is essential for workers to wear a face mask (which incorporates a special filter) at all times.

Metallic coating

In this process protection is provided by coating the components with a thin layer of metal having a greater resistance to corrosion than the component metal (e.g. aluminium, cadmium, nickel, zinc, tin, chromium). This will protect the component unless the coating is damaged. The protective coatings are applied to the component in a number of ways; electro-plating and dipping are the most common. Some processes are outlined below:

(a) *Hot-dip coating* can be used to coat the material with either a deposit of tin or a deposit of zinc. The application is most suitable for steel sheets and plates. The sheets or plates are passed through a bath of molten tin or zinc.

Since they are non-toxic, tin-coated materials (termed tin-plate) are used for food containers (e.g. canned food tins, milk cans), for domestic mincing machines and also for table ware. Zinc-coated materials (termed galvanized) are used for products such as buckets, dustbins, corrugated steel, water tanks, window frames, wheelbarrows and barbed wire.

A similar process is *sherardizing*, which also coats the component in zinc but uses powder instead of molten zinc. The components are packed into drums, together with zinc powder, and heated. During heating the drum is rotated to ensure all surfaces have contact with the powder and therefore become coated. Very close tolerances can be achieved by this process, which is used for coating items such as nuts and bolts.

(b) *Spraying*. The most common material sprayed is zinc but others, including aluminium, lead, tin, stainless steel, cadmium, copper and silver, can be applied. A special 'pistol' is used and heat (either by gas or electricity) melts the coating metal which becomes 'atomized' and is then sprayed on to a surface using a compressed air stream.

Since the process is portable and very flexible it has many applications which include the spraying of large structures such as bridges, electricity pylons and tanks.

(c) *Electro-plating*. An advantage of this process is that uniformity and thickness of the coating can be controlled much more accurately than with painting. The principle involved is to use the component which is being coated as the cathode in an electrolytic cell. The process is shown diagrammatically in Figure 194.

The anode is manufactured from the material to be deposited on the component. Both the anode (the positive electrode) and the cathode (the negative electrode) are immersed in a suitable liquid (called an electrolyte). When the electrodes are connected to an electrical supply and current allowed to pass through the circuit, molecules pass through the electrolyte from the anode to the cathode. As the anode dissolves, the cathode becomes coated due to the transfer of the molecules.

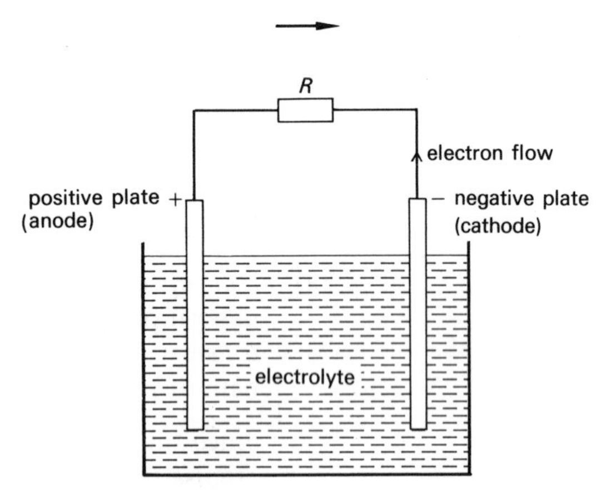

Figure 194 *A simple electrolytic cell. The anodes and cathodes may vary with different resulting reactions. The electrolyte must be compatible with both the anode and cathode*

Metals commonly deposited in this way are copper, cadmium, zinc, nickel, chromium, gold, silver and tin. Since no heating is required there is no danger to the structure of the parts being coated. The more widely used deposits are:

(i) *Chromium plating* gives a hard-wearing surface with a very bright finish, but is expensive. It is used most often for decorative purposes, e.g. on cars. In addition it is suitable for building up the worn parts of accurate measuring equipment such as gauges; the thickness of coating which can be deposited can be controlled within very small limits.

The thin deposits are usually coated on top of a nickel deposit. This gives a good adhesion.

(ii) *Nickel plating* is used primarily as a base for depositing a layer of chromium. It is hard wearing but not particularly resistant to industrial atmospheres.

(iii) *Zinc plating*. The application for which this is most suitable is as a surface preparation prior to painting. It is reasonably cheap. The plates or components being coated are hot-dipped (i.e. in molten zinc). The process is called galvanizing. Some uses are for cold water tanks, buckets, wheelbarrows.

(iv) *Cadmium plating* is a similar process to zinc-plating but provides much better protection, particularly where either moisture or salt are involved. Laundry equipment is usually cadmium plated but the process is not suitable for contact with foods (because of the poison risk) or for depositing on copper or brass.

Oxide coating

Materials such as aluminium readily combine with the oxygen in the atmosphere to form a corrosion-resistant oxide film on its surface. However, when aluminium is alloyed, the resultant material does not oxidize as readily. A process has been developed to produce an oxide film on materials such as aluminium alloys which make the alloy corrosion resistant. This process is called *anodizing*.

The work or component is suspended in an electrolyte and connected to the positive terminal. It is therefore the anode. The cathode can be either a separate electrode, usually of lead or stainless steel, or it can be the tank which holds the electrolyte.

The electrolyte is usually a sulphuric acid solution. When current is passed through the circuit atoms of oxygen are liberated at the surface of the component. These atoms combine with the aluminium and thicken the oxide film formed initially. The thickness of the film is usually controlled and varies between 0.007 mm and 0.015 mm. In addition to being corrosion resistant, the oxide formed, aluminium oxide, is also very hard and therefore wear resistant. Its surface is suitable for polishing and dyeing which produces a very durable and attractive finish.

For all methods of coating materials to prevent corrosion it is essential that the materials being coated are clean, dry and free from dirt, oil and grease. It is usual to subject the components to a cleaning operation prior to the coating. The cleaning processes and the plating processes involve the use of toxic materials. These materials must be handled and treated very carefully to prevent accidents (either due to contact with the materials or from fumes). Many of the materials are harmful and in some cases poisonous. The laws which govern the use and storage of such materials must be strictly obeyed.

Self-assessment questions

1 Name two dangers to health arising from paint spraying.

2 State three different methods of protecting surfaces from corrosion.

3 If unprotected steel is in contact with dry air, it corrodes.
TRUE/FALSE

4 Select from list B one application to which the form of protection of each process in list A is most suitable.

List A	*List B*
Zinc plating	(a) Base for chrome plating
Nickel plating	(b) Coating of laundry equipment
Anodizing	(c) Surface preparation prior to painting
	(d) Protection of aluminium and magnesium alloys
	(e) To provide a bright decorative finish

5 In electro-plating the components being plated are the *anodes*.
TRUE/FALSE

6 Why must some materials, particularly steel, be protected from the atmosphere?

7 Which metals listed below are suitable for protection by anodizing?
(a) Steel
(b) Magnesium alloys
(c) Zinc
(d) Chromium
(e) Cast iron
(f) Aluminium alloys

8 State three different methods of protection of surfaces by paint.

9 How can a fire hazard be caused when the stoving process is being used?

Topic area: Working in plastics

After reading the following material, the reader shall:

20 Appreciate the techniques available within a conventional workshop for producing shapes in plastic materials.

20.1 Describe the solvent and welding techniques of joining thermoplastics, indicating their applications.

20.2 Explain the use of heat-bending techniques for forming plastics.

20.3 Explain the use of casting techniques for forming plastics with special reference to encapsulating techniques for electrical components.

20.4 State the problems associated with the machining of plastics and the speeds and feeds necessary.

20.5 Explain the problems associated with producing holes by drilling in plastic due to clogging and overheating.

For many years apprentices have been trained in engineering skills using only metallic materials. However, in recent years the development and increasing use of plastic materials has ensured the introduction into training programmes of some of the techniques used to shape polymers. Many techniques are used in a conventional workshop. This text concentrates on the techniques most commonly used in industry. Some problems are similar to those encountered with metallic materials, but a number of basic processes must be specially adapted to suit plastic materials.

Joining of thermoplastic materials

Plastics, like metals, sometimes need to be joined to produce a complete component or assembly. Joining may be carried out in various ways. One method is to use adhesives (as described previously in the text). Other methods include:

(a) Joining using solvents
(b) Welding

Solvents. This form of joining is only suitable for thermoplastic materials. In this process, liquids are used which quickly dissolve the plastics being joined. The solvent also evaporates quickly during the production of the joint.

During the joining process it is essential to protect the component or material from contamination and damage from the solvent, apart from the localized area which is being joined. With the appropriate surfaces protected, the parts to be joined are immersed in the solvent. The solvent is usually placed in a tray which contains only sufficient depth of liquid to coat the unprotected areas. When the surfaces become 'tacky' they are removed from the solvent, positioned as required and then pressed firmly together. Pressure must be applied until the joint is dry.

The joint takes a few days to reach its maximum strength, so it is advisable to store the materials or components in a load-free position to prevent any distortion before service. With many plastics the drying time for the joint can be reduced by the application of low heat.

A second method using solvents is to dissolve some small pieces of thermoplastic material in the solvent. Provided that the proportions of each are correct, a paste is formed which can be applied as though it were an adhesive.

Typical applications include the mating surfaces of spigot and socket pipes and also lapped joints.

Solutions to self-assessment questions

1 Fumes from the paint which may overcome the operator. Breathing of the heavier paint particles which are difficult to remove from the chamber where spraying takes place.

2 Painting
Coating with a metallic surface
Protection by an oxide film

3 False. The air must contain moisture for the corrosion to commence.

4 Zinc plating – (c)
Nickel plating – (a)
Anodizing – (d)

5 False. They are the *cathodes*.

6 The atmosphere usually contains air and moisture, and these, in contact with a metal, cause an electric current to initiate corrosion.

7 (b) Magnesium alloys
(f) Aluminium alloys

8 Methods of painting include brushing, spraying, dipping, electro-static spraying, stoving.

9 The thinners in the paint evaporates quickly, creating inflammable fumes – a fire hazard.

Welding (i) – Hot-gas welding. This is a similar process to gas welding of metals (page 125). The gases used may be nitrogen or other inert gases.

Before commencing welding, the pieces being joined and also the filler rod must be cleaned (degreased). The edges to be joined must also be prepared. For a butt weld the edges may be square or chamfered to give an included angle of 60° when the two are placed in position. Sometimes, particularly with thicker materials, a double-vee is produced. Typical joint preparations are shown in Figure 195.

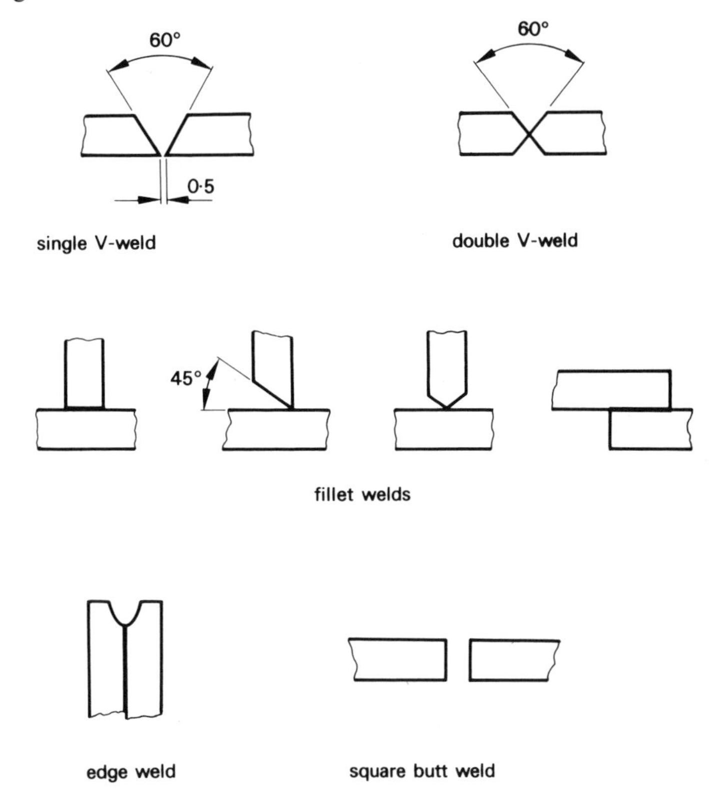

Figure 195 *Hot-gas welding of plastics*

The welding torch contains an element used for heating the gas to the approximate temperature for welding. A filler rod of the same plastic as that being welded is used; a common size is 3 mm diameter. To prevent movement of the pieces being welded, tack welds are used, or alternatively the pieces are firmly clamped to the welding bench. Tack welding is more suitable since it eliminates clamps which may interfere with the actual welding operation. Tack welding is carried out by using a special tack-welding tool which is attached to the gas torch. The filler rod is held almost vertically initially above the starting position, and the heat from the torch is directed to the bottom of the filler rod and the vee immediately below the rod. When the rod softens, the end should be pressed firmly into the vee to provide good contact between the rod and the parts being joined.

After contact is made, the torch is moved along the joint, still heating both sides of the chamfer and also the filler rod. The filler rod is now inclined at an angle varying between 45° and 60°, while pressure is continually applied to fill the vee with the filler rod and to form the joint (Figure 196(a)). Most processes of this type use a shoe to apply pressure to the rod/joint (Figure 196(b)).

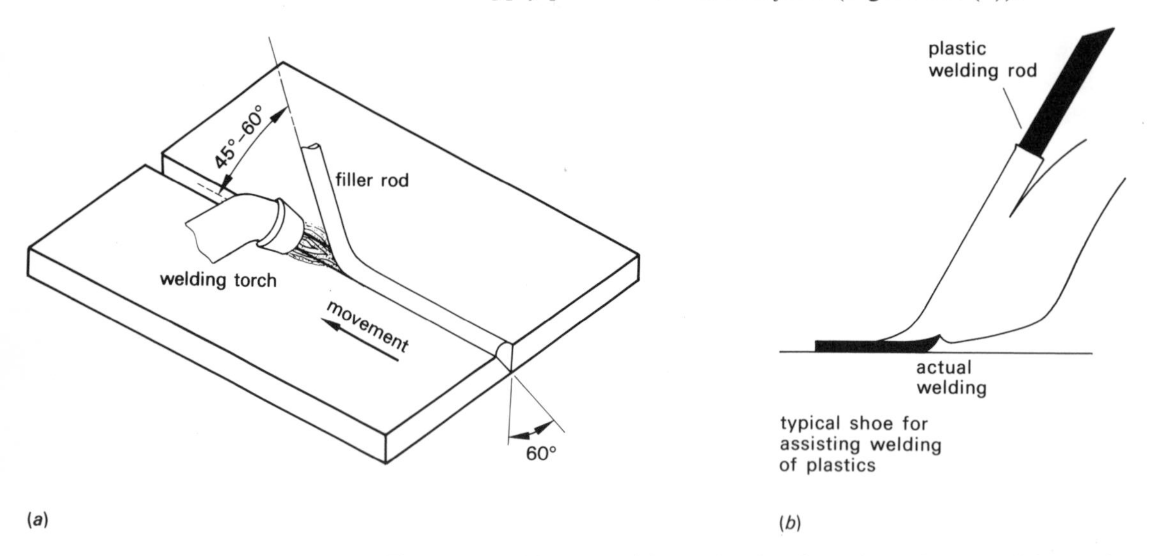

Figure 196 *Hot-gas welding – details. The rod may be rotated during the application of pressure to form a good joint*

Typical applications are very similar to the applications of gas welding of metals. Intricate work can be completed quickly and easily. It is suitable for ducting and tank liners.

Welding (ii) – Hot-tool welding. In this process the faces of the parts to be joined are heated separately but at the same time. Heated plates, tools or sockets which are thermostatically controlled are used. These soften the faces to be joined. When the faces are suitably soft the heat is removed and the faces pressed together. Pressure is removed when the joint has cooled. The process is often used for joining pipes, and is also very suitable for sheet and film. When joining sheet and film a tool similar to a soldering iron is used. Usually, when a butt joint is produced by this method, some material is displaced; this is illustrated diagrammatically in Figure 197, together with a representation of the process. The displacement of material is an indication of a good joint.

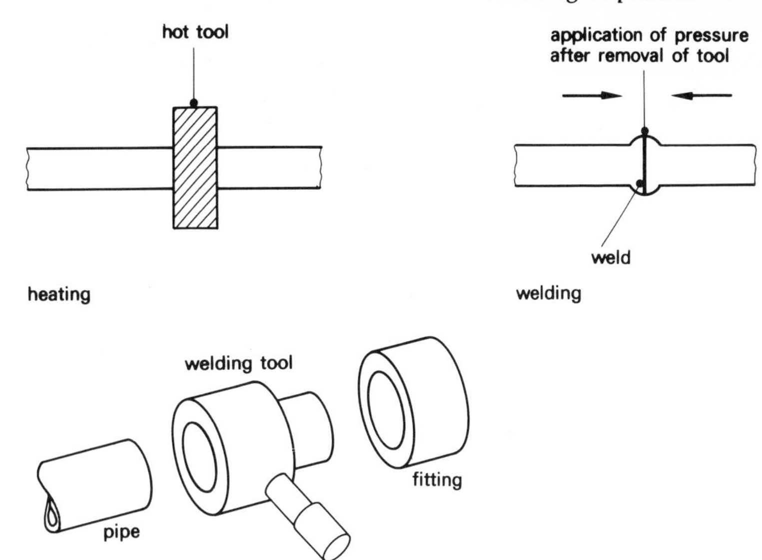

hot tool welding of pipes and fittings

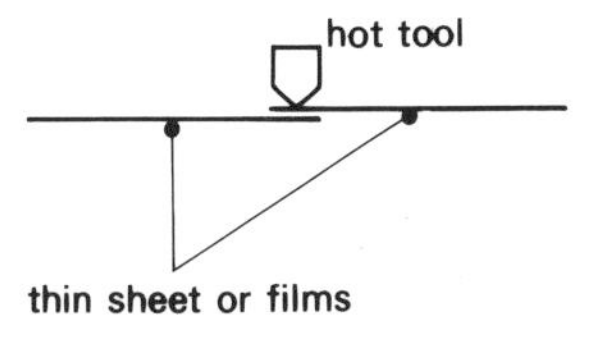

tool welding of thin sheet or films

Figure 197 *Hot-tool welding of plastics*

Typical applications include spigot and socket joints, butt joints, sheet materials and extrusions, liquid-filled sachets.

Welding (iii) – Friction welding. Whilst not being one of the most commonly used processes, friction welding can be carried out quite easily and effectively. Basically the method consists of positioning the components that are to be welded so that they are in contact. One of the components is held stationary while the second component rotates. A lathe or a drilling machine is suitable for this process. The rotating part is held in the chuck; the stationary component may be held in an attachment fitted to the tailstick of the lathe, or in a vice on the drilling machine. It is important that both components are accurately aligned.

When the component rotates at high speed, contact with the stationary component creates friction, which produces heat. The heat softens the contact areas of each component. When the contact areas are sufficiently soft, rotation is stopped and the parts pressed firmly together to produce a joint. Joints can be produced on both solid and hollow components, Figure 198.

The process is most suitable for joining circular parts, but new techniques enable non-circular parts to be joined. However, these techniques are not usually available in a conventional workshop.

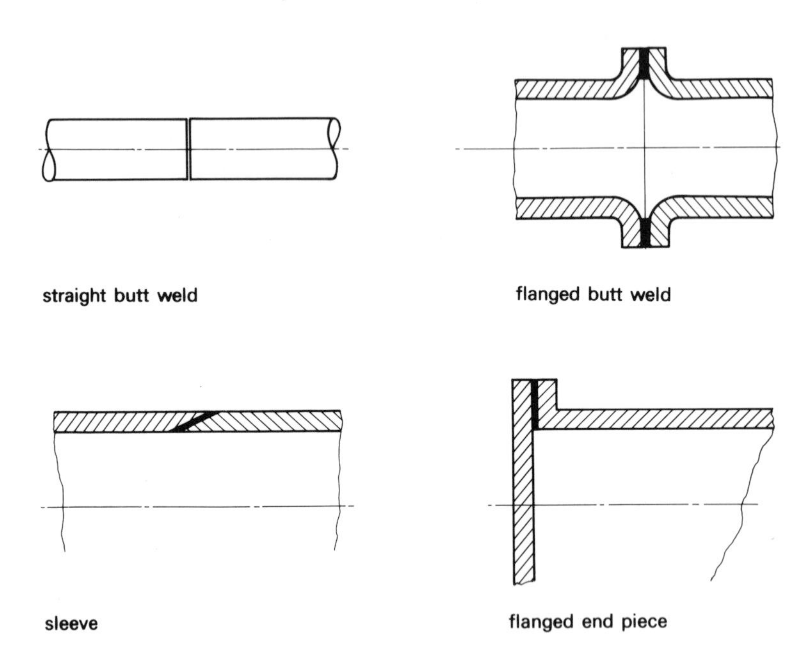

Figure 198 *Friction welding – typical joints. Many others are possible*

Dangers of joining methods

All methods of joining, whether by solvents or by welding, involve dangers. Some solvents may cause skin irritation; the fumes from others should not be inhaled. In the welding techniques, heat from components or flames may not be obvious, but can cause severe burns. Care must be taken in all instances.

Heat-bending techniques

In addition to welding, thermoplastic materials can be formed reasonably easily with equipment found in many workshops. The heat-bending process uses heat to soften a specific area of the plastic, and then bending is carried out to produce the desired form. The heat can be provided by an infra-red element (similar to types used in domestic electric fires). This can be fitted into a framework so that the area to be heated can be varied by adjustment of the plastic supporting boards. A typical arrangement is shown in Figure 199.

The support boards should be positioned to allow heat to soften only the area of the bend. Care should be taken to ensure that the plastic does not overheat, or it may begin to decompose. When bending is being carried out, it is advisable to use a jig or former so that the whole bend is uniform. Metal formers produce the best finish, since the grain from a wood former may affect the surface finish of the plastic. The minimum internal radius to which a piece of material is bent should not be less than the thickness of the material.

When the material has been bent to the desired shape, it should be held or clamped in the set position until it has hardened. The shape is then retained. As with other processes, the materials are hot, and care should be taken when handling to avoid burns.

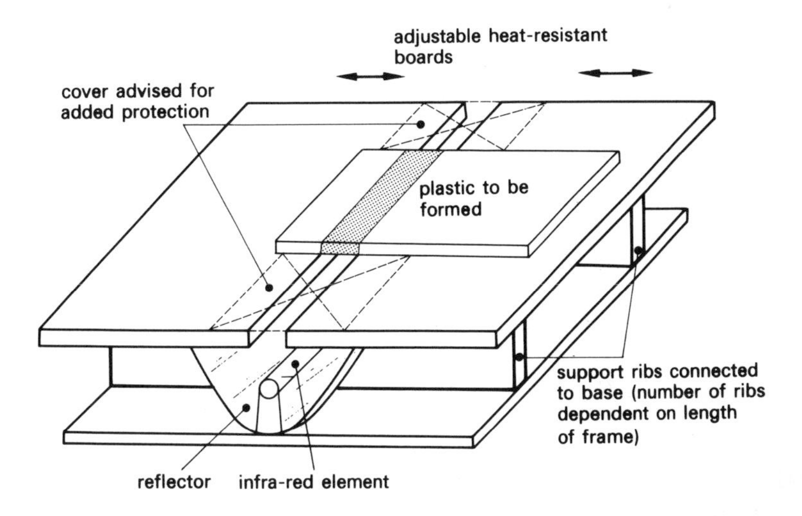

Figure 199 *Heat-bending equipment. Location can be adjusted to ensure correct positioning of heated area if several are to be produced*

The bending described so far is suitable for straight lengths of bends at varying angles and radii. Using more elaborate tools or formers in conjunction with standard machines in the workshop, it is possible to produce more complex shapes. Suitable machines are the lathe, drill or hand press depending upon the shape required. Rather than the local application of heat, the whole of the plastic may have to be softened. This can be carried out in a small oven or furnace. At the correct temperature the plastic is removed from the heat, placed between the formers and pressure applied to produce the desired shape. A typical example is shown in Figure 200.

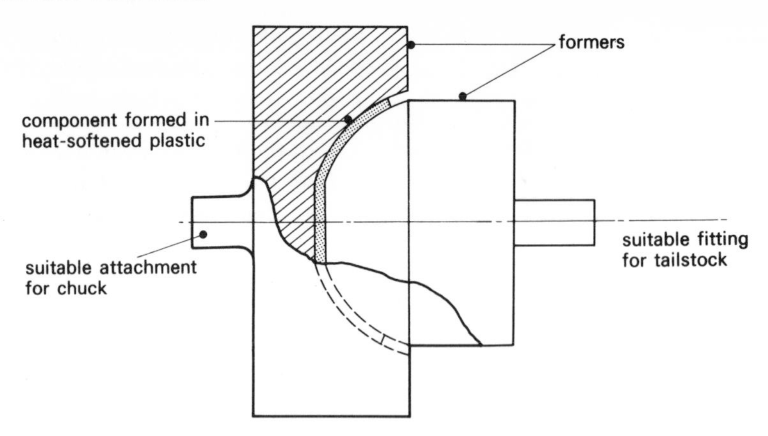

Figure 200 *Use of formers. Attachments may be modified to suit other equipment. An ejector may also be necessary for some shapes*

Casting techniques

The forming methods discussed previously are for thermoplastic materials. Thermosetting materials cannot be formed in these ways because during heat forming they become permanently hard. Some thermosetts are manufactured in liquid, uncured form, which can be poured into a mould to produce desired shapes. Some of the common thermosetts of this type are epoxy resins, phenolics and polyesters. Thermoplastics such as acrylic, polyethylene, PVC and urethane can also be cast, but the following discussion is concentrated on thermosetting materials.

The thermosetting material remains liquid until a hardener is added. The resin (the thermosett) and the hardener are mixed together, and poured into a mould. A chemical reaction takes place between the two materials with the result that the material, over a period of time, sets permanently hard. The mixture is then said to be cured. The thermosett may be cured at room temperature or (as is necessary for many materials) with heat. Curing can be accelerated by adding other elements which speed up the reaction.

The accuracy and surface finish of cast parts depend upon the care taken in the production of the moulds. There are two basic types of mould which can be used. One type is permanent, such that it can be used several times to produce the same or similar shapes, and the second is a temporary type of mould, suitable for making one casting only.

Permanent mould casting. These moulds can be assembled in wood, metal or plastic. A major problem when casting thermosetts is that they are liable to act like an adhesive and bond to the mould. To prevent this, a release agent must be applied either by spraying or painting all parts of the mould which make contact with the material being cast. Several release agents are available.

The mould may be made in one piece or a number of pieces, so that extraction of the casting is easy. After assembly of the pieces together with the coating of a release agent, the plastic is prepared for pouring. Usually the material is in two parts, the resin and the hardener. Only sufficient material is mixed to produce the casting; this prevents waste. When thoroughly mixed, the liquid is poured into the mould, spreading the material evenly within the mould. Care must be taken to prevent air being trapped during the pouring. When the mould is filled to the appropriate levels, it is allowed to cure. This may be at room temperature, or the mould can be placed in an oven at a specific temperature. The temperature used and the time to cure depend very much on the manufacturer's instructions. Before removing the casting from the mould it is essential to ensure that it has fully cured and that it is not 'tacky'.

Some of the possible forms, showing typical moulds, are shown in Figure 201.

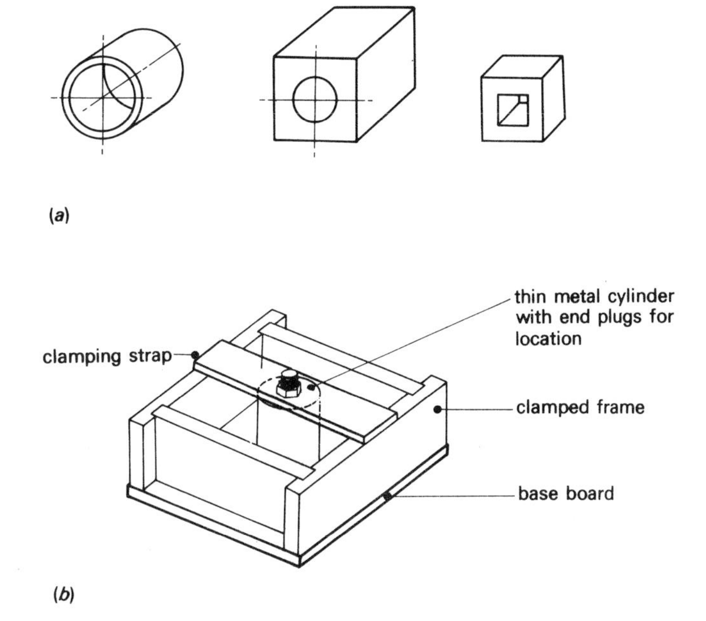

Figure 201
(a) *Some typical shapes for casting; many others are possible (including solid forms)*
(b) *Typical mould for casting plastic material*

Temporary mould casting. More complex shapes than those shown in Figure 201 may be difficult to extract from permanent moulds. To overcome this difficulty a temporary mould is produced. The mould is usually made from plaster; after casting, the mould is broken in order to remove the component. If a temporary mould is to be used, a pattern is required. The pattern may be made in wood, and is the same general shape of the piece to be cast. There may be slight differences between the pattern and the product being cast. If the product has square sides, draft angles are included on the pattern; this enables the pattern to be withdrawn from the mould far more easily and with less chance of damage to the mould. Allowances may also be included to offset any possible shrinkage on solidification. A third difference may be an allowance for machining. The procedure for producing the mould from the pattern is shown in Figure 202.

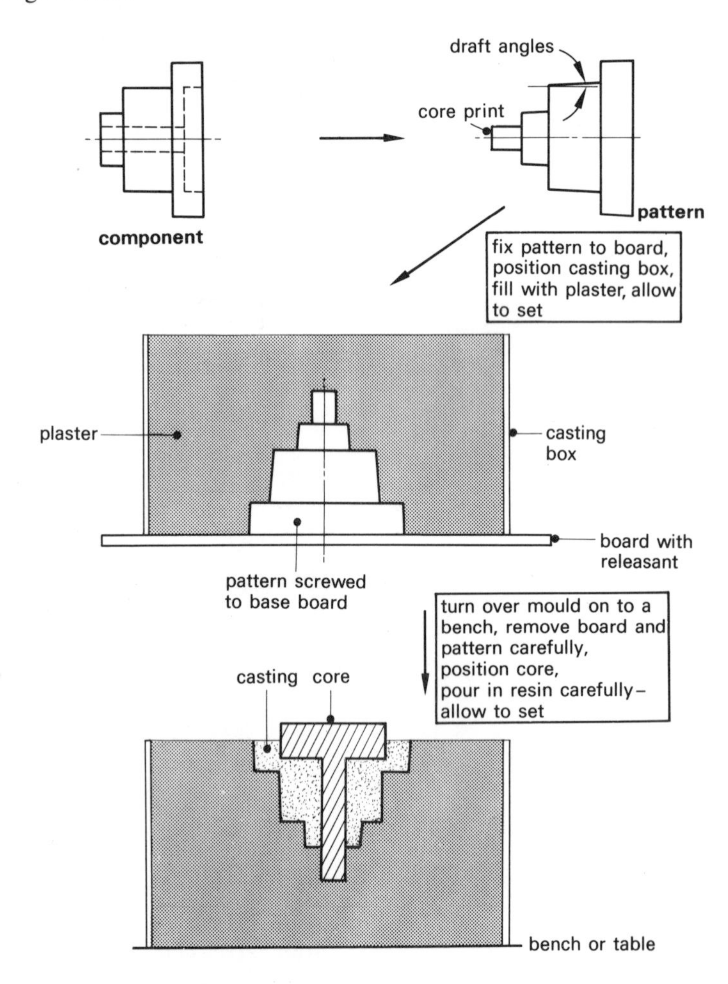

Figure 202 *Making a temporary mould*

When the mould is ready, the procedure for producing the casting is exactly the same as that described previously. For each part produced a new mould must be made.

Another use for the casting process is to produce a block in which an object is encased. These are typical of some of the paper weights which are currently produced. A block of this type has to be produced in stages. The method is shown in Figure 203.

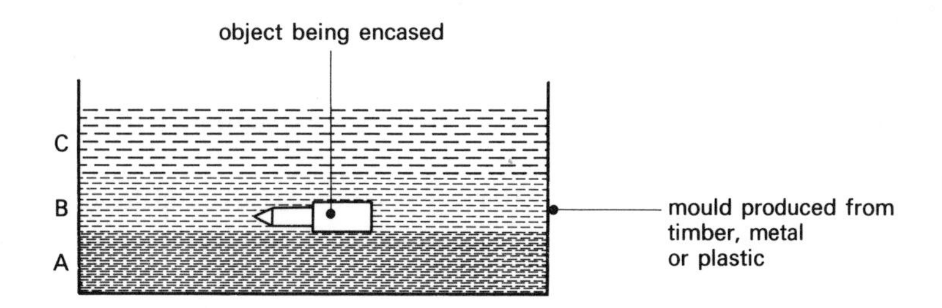

Figure 203 *Encasing in plastic*
(i) *Layer A is poured and allowed to set*
(ii) *The object is placed in position on top of layer A*
(iii) *Layer B is then poured around the object and allowed to set*
(iv) *Layer C may be clear (as A and B) or it may have colour added for effect*

A variation of this procedure is to cast a handle for a tool. This can be done either as a separate handle which is fitted as required on to the tool, e.g. a file, or cast directly on to the tang of the tool as a permanent fixture. In practice these are normally injection moulded and not cast.

Encapsulating of electrical parts

The casting technique is widely used for encapsulating electrical parts. The encapsulating creates the conditions necessary for satisfactory functioning of the parts. These conditions include protection against vibration, dust, moisture or other contamination, and provision or improvement of insulation. Encapsulating provides a stable environment in which an electrical component can function efficiently.

The encapsulating can be for one part only, or for several which are assembled together, e.g. a small circuit. As technology has advanced, electrical and electronic parts have become considerably smaller, and therefore difficulties in assembling components have increased, as has positioning. This has been overcome by encapsulating.

The process can be used to protect joints on cables but is particularly used for encapsulating capacitors, diodes and coils. A more common name for the process is 'potting'.

Machining of plastics

Most plastics can be formed easily to produce components with a good surface finish. Sometimes it is desirable to machine the component after forming.

The reasons for machining include:

(a) A greater accuracy is required than is normally achieved during forming

(b) The complexity of the component is such that it is more economic to machine after forming, rather than to manufacture solely with forming processes.

Laminated or reinforced plastics cannot be formed in the usual way, so machining is often essential.

Plastics can usually be machined by use of conventional wood- and metal-working machinery, but small alterations are necessary to the tooling. The most important consideration is the generation of heat during the machining. Plastics are much more heat-sensitive than metals, compared with metals they have a considerably lower thermal conductivity, but the coefficient of thermal expansion is much greater. To offset these problems, coolant may be used; also light cuts and low feeds are necessary. In addition, cutting speeds should be high.

The cooling medium may be water and soluble oil, or air if this is available. A general guide to the common machining processes is given below.

Turning

Many plastics can be turned using a high-speed steel tool at speeds ranging from 90 – 180 m /min. For longer runs or for glass reinforced plastics and thermosetts, a carbide-tipped tool is better. The tools should be kept sharp, and it is preferable to use a round-nose rather than single-point tool. Typical feeds, speeds and angles are summarized in Figure 204.

Screw cutting can be carried out, but threads having a rounded form are produced more easily than threads of square form. Self-opening die heads combined with thread chasers should be used, ensuring that the thread is correctly cut. It is important to use light starting and finishing cuts to obtain a good finish.

material	*top rake*	*front clearance*	*feed* (mm/rev.)	*cutting speed* (m/min.)
polyethylene, PTFE	0° to −5°	18° – 30°	0.1 – 0.25	90 – 150
nylon, polypropylene, acetal	0° to −10°	18° – 30°	0.1 – 0.25	150 – 300
PVC (rigid)	0° to −10°	15° – 25°	0.2 – 0.6	90 – 300
acrylics	0° to −5°	15° – 20°	0.125 – 0.25	30 – 150
polystyrene	0°	15°	0.05 – 0.2	90 – 300
thermosetts	0° to 15°	15°	0.125 – 0.25	75 – 480
phenolics	0° to 30°	12° – 15°	0.125 – 0.25	180
glass-reinforced plastics	15° to −5°	0° – 15°	0.1 – 0.2	90 – 150

Figure 204 *Cutting speeds for plastics*

Drilling

Any type of drilling machine can be used for drilling plastics; whatever type is used, the main concern is to prevent the heat generated from affecting the material. If this is not done the material becomes tacky and clogs the drill, filling the flute instead of flowing away. In addition, the drill should be cleared frequently of swarf; this then prevents clogging which may generate unwanted heat.

The standard type of high-speed twist drill is suitable, but the rake and the clearance angles must be ground in accordance with the figures suggested for turning. The normal 118° point-angle drill is suitable for many plastics, but for materials such as nylon, polyethylene and polypropylene a point angle of 80 – 90° is more suitable. A problem may arise when drilling thin materials; if care is not taken damage may occur as the drill breaks through the bottom surface. Where possible, it is advisable to use a drill with a point angle large enough to ensure that the full diameter of the drill enters the hole before the point breaks through the bottom surface. This may mean point angles of about 140 – 150°. In addition, when drilling thin materials a backing board should be used; this helps to prevent cracking as the point of the drill breaks through.

As with turning, carbide-tipped drills are used for production runs, and also when drilling glass-reinforced materials. Since plastics are resilient, holes are drilled slightly oversize (0.05 – 0.125 mm) to maintain accuracy of the hole. If a hole of accurate diameter is required, then, as with metals, it is advisable to drill undersize and ream to the desired size. Feed rates for drilling are of the order of 0.15 – 0.20 mm per revolution.

After drilling, holes may be tapped either by hand or by machine. Once again, threads with a rounded form are produced more satisfactorily than threads of standard form. Slower cutting speeds than for drilling are used. Typical tapping speeds for sizes up to 12.5 mm are 200 revs/min. Withdrawal and clearing the tap are three or four times greater. Some typical drilling speeds are given in Figure 205.

In addition to turning and drilling, plastics can be formed by sawing, milling, shaping and grinding. Grinding is particularly suitable for materials such as the thermosetts and the filled or reinforced materials.

material	*speed* (rev./min.)		
	Ø 3 mm	Ø 12 mm	Ø 25 mm
polyethylene, PTFE	1000	500	300
nylon, polypropylene, acetal	1200	800	500
PVC (rigid)	5000	1000	500
acrylics	6000	1000	600
polystyrene	1500	700	300
thermosetts	8000	1500	600
phenolics	8000	2000	700
glass-reinforced plastics	1000	500	300

Figure 205 *Drilling speeds for plastics*

Self-assessment questions

1 State three different forms of welding used for joining plastic materials.

2 Describe how plastics are joined by the use of solvents.

3 Select the joining processes in list B below which are appropriate to the materials in list A.

A	*B*
thermosett	solvents
thermoplastic	adhesives
	welding

4 For each joining process listed below suggest a suitable application:
(a) Solvents
(b) Hot-tool welding

5 Thermosetting plastics can be formed reasonably easily in the workshop by the heat-bending technique. TRUE/FALSE

6 At what stage during the heat-bending process is the plastic removed from the heat in order to be formed?

7 After forming by heat-bending techniques, what must be done to the component to ensure retention of the formed shape?

8 Name two machine tools which are found in the conventional workshop that can be used for forming plastics whilst hot.

9 For which material, thermoplastic or thermosett, is the casting process particularly suitable?

10 Describe the process of casting a thermosetting plastic.

11 Why must permanent moulds be sprayed or coated with special materials before casting?

12 For which application has the casting process been widely adopted?

13 Plastics are usually machined on conventional machines using tools identical to those used for cutting wood and metal.
TRUE/FALSE

14 State three problems associated with machining plastics.

15 Choose from the list below the appropriate conditions for machining plastics.
(a) Very low cutting speeds
(b) High feeds
(c) Light cuts
(d) Low feeds
(e) High cutting speeds
(f) Single-pointed tools
(g) Heavy cuts
(h) Round-nose tools

16 For each of the conditions in list A below select the appropriate tool from list B.

A	*B*
Glass-filled plastics	High-speed steel tool
High production of thermoplastics	Carbide-tipped tool
Thermosetts	

17 When drilling thin plastic sheets, how is cracking of the sheets prevented when the drill point breaks through?

Solutions to self-assessment questions

1 Hot-air/gas welding; heated tool welding; friction welding.

2 The plastic surfaces to be joined are dipped in an appropriate solvent until they become tacky. At this stage they are removed and pressed together. Pressure is applied until the joint is dry.

3 All the processes in list B are appropriate to thermoplastic materials. Only adhesives are suitable for thermosetting plastics.

4 Typical applications of solvents include lapped joints and the surfaces of sockets and spigots.
Typical applications of hot-tool welding include sheets and films, butt joints, extrusions and liquid-filled sachets.

5 False.

6 Heat is applied until the plastic softens. It is then removed and formed.

7 The formed component must be held clamped in position until it has cooled.

8 Machine tools which can be used for forming plastics are the lathe, drill and fly press.

9 The casting process is particularly suitable for thermosetting plastics.

10 A mould, which may be permanent or temporary, is prepared, and the liquid thermosetting plastic is made ready for pouring. The liquid contains a hardener. Over a period of time the liquid sets hard to form the shape of the cavity in the mould.

11 The special coating prevents the cast material sticking to the mould; many thermosetts have adhesive properties.

12 A very common application of the casting process is the encapsulating of electrical components.

13 False; whilst the machines are the same, the tools have to be modified to suit the materials to be cut.

14 Some of the main problems associated with machining of plastics are:
(i) Plastics are much more heat-sensitive than metals.
(ii) They have very low thermal conductivity.
(iii) They have high thermal expansion.
(iv) Drills in particular are liable to clogging.

15 The appropriate conditions for machining plastics are:
(c) Light cuts
(d) Low feeds
(e) High cutting speeds
(h) Round-nose tools

16 The most appropriate tool is a carbide-tipped tool for all of the materials in list A.

17 The point angle of the drill is ground so that the whole of the drill diameter enters the material before the point breaks through. A support board is used beneath the plastic sheet.

Index